Higher Education Policy for Tackling Climate Change

Jouni Kekäle • Rómulo Pinheiro
Editors

Higher Education Policy for Tackling Climate Change

Drivers, Dynamics, and Effects

palgrave
macmillan

Editors
Jouni Kekäle
University of Eastern Finland
Joensuu, Finland

Rómulo Pinheiro
Department of Political Science
University of Agder
Kristiansand, Norway

ISBN 978-3-031-78478-1 ISBN 978-3-031-78479-8 (eBook)
https://doi.org/10.1007/978-3-031-78479-8

This Palgrave Macmillan imprint is published by the registered company Springer Nature Switzerland AG.
The registered company address is: Gewerbestrasse 11, 6330 Cham, Switzerland

In Memoriam

In loving memory of our dear colleague Jan Kohoutek (1978–2024).

Contents

Notes on Contributors

Dominik Antonowicz is Chair of Department of Science and Higher Education Studies, Nicolaus Copernicus University, Poland. He also worked as a postdoc scholar at CHEPS, Twente University, CIPES at University of Porto, and OISE at University of Toronto. He specializes in policy analysis including agenda-setting, implementation, and evaluation studies with a particular focus on government–university relationships, institutional governance, and leadership.

Justin Axel-Berg has worked as research coordinator of Projeto Métricas (FAPESP), based at the Faculty of Economics and Administration at the University of São Paulo (FEA/USP), Brazil, since 2017. He is a researcher in university governance and higher education development, with special interest in Latin America, Eastern Europe, and Central Asia. He has a master's degree in international relations (IRI-USP) and bachelor's in philosophy from the University of Sussex and Monash University. He is a member of the International Triple Helix Association.

Elizabeth Balbachevsky is an associate professor at the Department of Political Science at the University of São Paulo, Brazil, and Deputy Director of the University of São Paulo's Center for Public Policy Research. She is a fellow at the Higher Education Group (LEES), at the State University of Campinas (UNICAMP). She was the Brazilian Ministry of Education's General Coordinator of policies supporting HE internationalization between 2016 and 2017, a Fullbright New Century Scholar for 2005–2006. She is associated with HEG, Higher Education Group,

University of Tampere, Finland. Elizabeth Balbachevsky's main research interests is higher education policy.

He Botao is a master's degree candidate at the Institute of International and Comparative Education, Beijing Normal University, China. His research interest is on environmental education.

Anna Budzanowska is Doctor of political sciences at the Silesian University in Katowice, Poland. She is a specialist in the history of political doctrines and public policy, combining her academic career with her work in public administration. Since COP Katowice 2018 to date, she has been the coordinator of the Polish panel at the Climate Law and Governance Day 2021, a networking group that is part of the UN COP conferences.

David Charles is Professor of enterprise and innovation at Newcastle Business School, Northumbria University in the UK. His research interests are focused on the roles of universities in their regional contexts, regional innovation policy, clusters, knowledge exchange, innovation management, and urban development. He has led research projects funded by the UK Economic and Social Research Council, the European Commission, OECD, UK national government departments, and various regional and local organisations. He has published in a range of international journals including Regional Studies, Research Policy, European Planning Studies, and Science and Public Policy.

Evandro Coggo Cristofoletti is a post-doctoral researcher in Public Administration at the School of Applied Sciences of University of Campinas, Brazil. Evandro's research interests, from Public Policy Studies and Science and Technology Studies, are in the relationship between research, university, and society.

Alina Felder is a postdoctoral researcher at the School of Economics and Political Science at the University of St. Gallen, Switzerland. She obtained her PhD from the University of Bamberg, Germany. Her research interests include the governance of (EU) public policy, the role of universities in policymaking, and Europeanisation processes. While her research has appeared in journals including Contemporary European Politics and Politics and Governance, her monograph on the territorial dimension of EU knowledge policies is forthcoming with Routledge. Alina is a Steering Group member of the European Consortium for Political Research (ECPR) Standing Group on "Knowledge Politics and Policies".

Andrew Gibson is Assistant Professor in Philosophy of Education, at the School of Education, Trinity College Dublin, where he is also co-director of the Cultures, Academic Values in Education (CAVE) research centre. His interests are within the philosophy of higher education and critical higher education studies. He is a member of the executive board of the Philosophy and Theory of Higher Education Society (PaTHES), international advisory board of the Centre for Higher Education Futures (CHEF) in Aarhus University, the editorial board of *Philosophy and Theory in Higher Education*, and is reviews editor for *LATISS Learning and Teaching—The International Journal of Higher Education in the Social Sciences.*

Ellen Hazelkorn is Professor Emeritus, Technological University Dublin TU Dublin, Ireland. She is also Joint Managing Partner at BH Associates and Joint Editor for Policy Reviews in Higher Education. She has worked with/advised international organisations and governments as well as universities for over 20 years, including the European Commission, OECD, World Bank and UNESCO, and was Vice President, Dublin Institute ofTechnology (now TU Dublin). Ellen has published almost 200 peer-reviewed articles, book chapters, short articles and blogs, commentaries, and policy papers including on higher education and civic/regional engagement.

Jouni Kekäle is Professor and Chief Senior Specialist working at the University of Eastern Finland. He has previously acted as Human Resources Director in multidisciplinary universities of Joensuu and Eastern Finland for 23 consecutive years. His main research interests have included academic leadership, human resources management, quality, responsibility, disciplinary and organizational cultures in Higher Education, as well as assessment systems and the changing operational environment for the higher education sector. He was a PI in Horizon2020 sub-projects on researcher assessment and now leads an EU-funded task Recognition and Rewards in YUFE alliance, a collaboration between ten European institutions.

Jan Kohoutek (1978–2024) was an Associate Professor of the Department of Public and Social Policy at the Faculty of Social Sciences, Charles University, the Czech Republic. Holding a PhD in public policy, in the higher education field, he researches into university governance, third mission policies, inbreeding or quality assurance. On these topics, Jan has published in peer-reviewed international journals, including *Minerva*, *Higher Education Policy*, and *Higher Education Quarterly.*

Tao Linlin is a researcher who got her Master of Education from the Institute of International and Comparative Education, Beijing Normal University, China. Her primary research area focuses on climate change education and nature-based education, in which she has worked on several projects and has completed several papers and book chapters.

Jacques Marcovitch is a Professor at the University of São Paulo (USP), Brazil. He studies business pioneering, strategy, and innovation with a focus on economic growth, income distribution, and environmental sustainability. Since 2002, he has been researching policies to implement the Climate Convention, with an emphasis on reducing greenhouse gases. He was Rector of USP from 1997 to 2001, Pro-Rector of Culture and University Extension at USP from 1994 to 1997, formerly director of the Institute for Advanced Studies and FEA/USP.

Rómulo Pinheiro is Professor of Public Policy and Administration, Deputy Head of Department of Political Science and Management and member of the research group on Public Governance and Leadership (GOLEP) at Norway's University of Agder (UiA). He is also a member and co-track coordinator of the Center for Digital Transformation (CeDiT) based at UiA. Rómulo's research interests are located at the intersection of public policy and administration, organisational studies, economic geography, innovation, and higher education studies. Topics include: mergers, resilience, change and adaptation, institutionalisation, societal engagement/impact, culture and identity, digitalisation, management and governance. Empirically, the focus is on the Nordic countries (comparative) and the higher education sector (universities and science systems as institutions).

Emanuela Reale is Associate Senior Researcher and former Director at the Research Institute on Sustainable Economic Growth—National Research Council (CNR-IRCRES). Her areas of interest are higher education policy, higher education governance, R&D funding, STI indicators, research evaluation and impact assessment—with a special focus on SSH research. She has been PI or Coordinator of several European projects such as Prime-NoE, JOREP, PREF, and RISIS. She is President of the European Forum for Studies on Policies for Research and Innovation—Eu-SPRI. Emanuela is co-editor of Research Evaluation and published in many international Journals dealing with research policy, research evaluation, and higher education studies.

Andrea Orazio Spinello is a researcher at the Research Institute on Sustainable Economic Growth of the National Research Council of Italy (CNR-IRCrES). His research focuses on research policies and evaluation, with major attention to public investments in R&D and changes in scientific practice at the individual level. He is one of the leaders of a nationally relevant research project on universities' community engagement. He is involved in projects aimed at developing and maintaining research infrastructures for social sciences. He is one of the managers of the EFIL dataset, which is part of RISIS, a European research infrastructure for STI studies.

Joanna Suchomska is a sociologist associated with the Department of Human Ecology at the Institute of Sociology at the Nicolaus Copernicus University in Torun, Poland. She is a specialist in urban sociology, environmental sociology, and participatory governance. In cooperation with the Sustainable Development Laboratory (NGO), she works with city authorities to develop and implement sustainable development policies at the local level.

Michał Wróblewski is a sociologist and a philosopher. He is acting as director of the Institute of Sociology at Nicolaus Copernicus University in Toruń, Poland. He is interested in science and technology studies, sociology of medicine, health and illness, environmental sociology. His current research is focused on vaccine hesitancy and environmental activism.

Huang Yu is an associate professor at the Institute of International and Comparative Education at Beijing Normal University, China. His research interests encompass environmental education, education for sustainable development, and geography education. He has authored around 70 articles and written more than 30 books. His professional career outside of academia includes writing and editing textbooks and popular science books for a variety of publishers. He is a member of both the Geographical Society of China and the Chinese Society of Comparative Education.

List of Figures

List of Tables

CHAPTER 1

Climate Change, Higher Education and the Wicked Problem of Time

Jouni Kekäle and Rómulo Pinheiro

Introductory Background

In this edited international volume, we investigate the role of universities and other types of higher education institutions (HEIs) in tackling climate change (CC) in the context of political steering and policy frameworks. Whereas earlier work was largely focused on broader global issues related to sustainability, most notably insofar the importance attributed to the United Nation's Sustainable Development Goals (SDGs), less attention has been paid to the policy framework (central and local governments) with respect to the role of higher education (HE) systems and HEIs in CC.

J. Kekäle (✉)
School of Educational Sciences and Psychology, University of Eastern Finland, Joensuu, Finland
e-mail: jouni.kekale@uef.fi

R. Pinheiro
Department of Political Science and Management, University of Agder, Kristiansand, Norway
e-mail: romulo.m.pinheiro@uia.no

J. Kekäle, R. Pinheiro (eds.), *Higher Education Policy for Tackling Climate Change*, https://doi.org/10.1007/978-3-031-78479-8_1

This edited volume deals with the HEIs' and HE policies' roles with respect to CC mitigation through different country cases (to be introduced at the end of this chapter in more detail). This way we can provide multiple international analyses, with an overall comparison in the final chapter of the volume. In this volume, scholars from the fields of higher education studies have analysed HEIs' role in CC mitigation and the steering policies affecting these matters in diverse national contexts such as Central and Eastern Europe (Poland, the Czech Republic, Switzerland and the Netherlands), Southern Europe (Italy), the UK and Ireland, in addition to leading economies in world regions such as Asia (China) and South America (Brazil).

In this introductory chapter, we first map the big picture of the global problem of CC, what causes it and the key players in the area. The discussion is somewhat simplified but hopefully gives insights of why the problem is pervasive and what could be HEIs' and policy's potential roles in global CC mitigation. We then turn into the policy—HEI nexus and discuss the shifting overall policies and their steering effect on the HE sector. Short descriptions of the empirical case studies conclude the introduction.

Climate Change

Climate change is arguably the most crucial problem of our era, and perhaps the grandest challenge humanity has yet faced. Cook et al. (2013) have noted that 97% of scientists concur that climate change is real and is caused by human action. The problem is tightly connected to our modern ways of life. Climate change is caused by societies' excessive emissions, which cannot be absorbed by the earth's natural systems. Burning fossil fuels causes emissions in the form of carbon dioxide (CO_2), exceeding the amounts that may be handled by plants in photosynthesis or absorbed by sinks in forests, soil and the seas. Excessive carbon dioxide remains in the atmosphere for hundreds or thousands of years. Methane is a much stronger greenhouse gas but remains in the atmosphere only around eight years.[1] Anthropogenic methane production is connected to the

[1] Nitrous oxide (N_2O), laughing gas, is a strong greenhouse gas causing much of agriculture's emissions which account for 16–27% of human-caused climate-warming effect. Synthetic nitrogen fertiliser is the main source of N_2O. Nitrous oxide is around 300 times as potent as CO_2 in warming climate, and typically remains in the atmosphere over one hundred years before disintegrating (Chrobak, 2021).

production of gas and coal (Taalas, 2021), the major energy sources from the onset of the first industrial revolution in the mid-eighteenth century. These so-called greenhouse gasses warm the atmosphere by absorbing the sun's heat that radiates from the earth's surface, preventing it from escaping into space.

In 2016 around 70% of CO_2 emissions emanated from the energy sector (petroleum, oil, coal and gas), an industry involved in the production and distribution of energy required for the production and transportation of goods and services in modern economies (Ritchie, 2020, as shown in Fig. 1.1). Notably, this industry includes the burning of oil and coal to produce energy for electricity, heat, industrial manufacturing and transportation. The production of electricity and heat plays the largest role at the global level, with close to 25% of all emissions. These are global measures, but the numbers also vary locally: in the United States, transportation is a much larger contributor of emissions than the global average, and in Brazil, most emissions come from agriculture and land use (Ritchie et al., 2020) (Fig. 1.1).

There is an imminent need for cheap green energy but an even more pressing and everyday need for reliable and affordable energy for all citizens. At present, the cheapest and easiest way to provide energy at scale is

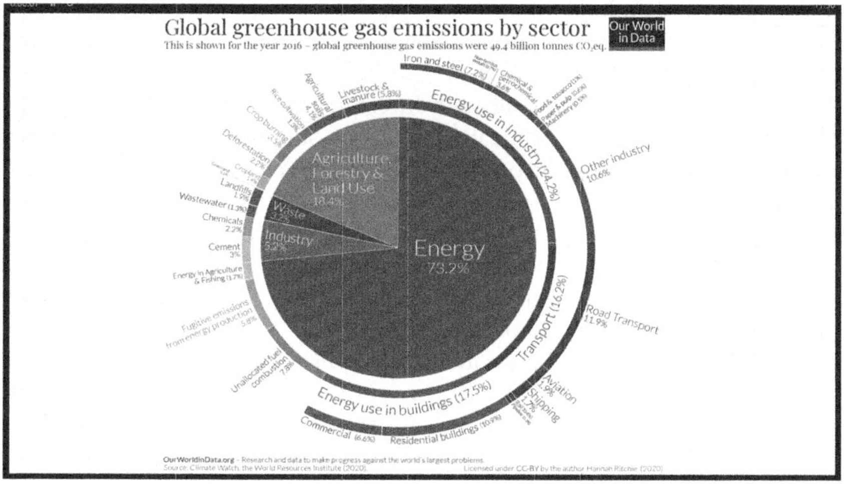

Fig. 1.1 Global emissions by sector (2016). Source: Ritchie (2020)

through fossil fuels. Green alternatives, like wind energy, are on the rise (Global Offshore Wind Report, 2022; Masterson, 2022), and have, in some jurisdictions, become the cheapest alternative. However, green energy production is largely dependent on weather conditions. Around 80% of global energy still comes from fossil sources. Supranational bodies such as the European Union (EU), in the context of its Green Deal, are investing heavily in green transitions (Hainsch et al., 2021), but there is a recent awareness that the set targets (volume and timeframes) are too ambitious. In addition, post-COVID19 economic realities and geo-political tensions (Ukraine and the Middle East, to name a few), alongside the rise of right-wing populism, have led to a wake-up call and a scaling back of ambitions in recent years (cf. Lockwood & Lockwood, 2022).

A steadily growing number of HEIs across the world are playing a leading role in the fight for a sustainable world and against CC. Studies show that relevant work is being done at different levels and across different types of universities and other types of tertiary education institutions on these matters (Leal Filho et al., 2019; McCowan et al., 2021). However, HEIs do not act in isolation. To understand HEIs' potential roles in CC mitigation (especially when keeping in mind the energy business) we need to adopt a broader (more holistic and systemic) perspective that considers the complex interplay between key drivers and mechanisms at the macro (policy), meso (HEIs) and micro (individual agents) levels. There is also a larger system of influential actors at play. The 'wicked problem' (Head, 2022) is so difficult to answer because of complex interactions and interdependencies.

Empirically investigating all such interactions would be too complex a task but we can, as a theoretical orientation, simplify matters and discuss this complexity as a background for the question of HEIs' and governmental policy's potential roles in CC mitigation. Thus, although beyond the scope of our research mandate, in addition to the policy-HEI nexus, investors, companies and citizens play a crucial role especially when we keep in mind the key questions related to energy sector's emissions which are fundamental for the whole problem of CC. Understanding the phenomena of CC and HEIs' potential role requires expanding the discussion to this larger system, composed of a multiplicity of nested sub-systems, economy, politics, culture, society, etc. We attempt to provide a concise but simplified snapshot of the contours (actors, dynamics and issues) of such a system below.

Today it is broadly agreed that the ultimate solution to CC is to get rid of greenhouse gasses (Gates, 2021; COP 28). Another basic fact is that emissions from the energy sector is the key issue in global CC. That problem would need to be dealt with in any successful major CC mitigation programme. In understanding HEIs' role in CC mitigation, it is evident that there are also other key operators affecting the fundamental energy sector as well as national and supranational (e.g. EU) policy towards CC more broadly. To get an initial understanding of the complexity of the problem, we deal with interactions between key players: major companies who run the global energy business; politicians who have the right to regulate societies and who steer public expenditures; HEIs who are instrumental in socialising future citizens and producing basic science resulting in current energy solutions—and in developing new solutions; and communities and the public at large as consumers making the demand for energy.

Climate Change and Green Transition: The Stakeholders' Dilemma

When considering the fundamental energy sector, a simplified picture is that there are four key players or stakeholder groups which currently see CC as an urgent issue but cannot really tackle it on their own. The key players all have their own typical time spans, pressures, drivers and inertias. A simplified understanding of the broad system reads as follows:

Politicians and *policymakers* (local, national and supra-national levels) control public resources and have the right to regulate green transitions. No one else can do that. They have other urgent and conflicting issues and pressures on the table as well, most notably job creation/economic growth, defence, immigration, declining birth rates, etc. Moreover, their operational (attention) span tends to be relatively *short*. Elected politicians tend to avoid any unpopular and bold decisions, particularly with respect to long-term dilemmas such as CC, as re-elections are always around the corner. This has been famously described in a quote by Jean-Claude Juncker: "We all know what to do, but we don't know how to get re-elected once we have done it" (Goodreads, 2024). Climate change is even trickier, as there appears to be no direct and easy solution to be implemented to eradicate the fossil fuels which are so intertwined with our modern way of life.

HEIs have (in principle) time to consider and study CC in the long term but they do not have unlimited resources to invest in green energy nor the right to regulate CC issues and policy in society. HEIs' highest potential is in producing new knowledge, innovation and education. Innovations centred on fostering and advancing the utilisation of cheap green energy would be among the most urgent contributions (Gates, 2021). Some basic, technological solutions exist but there is still a need for improvements and innovations to make such solutions scalable, affordable and secure. HEIs' primary activities of teaching and learning (induction of skills, competencies and attitudes) and basic and applied research (knowledge production and transmission) are, in the eyes of many, critical functions in supporting world societies in tackling mitigation targets.

Globally, HEIs are increasingly pressured to enhance the social impact of their teaching, research, innovation and engagement activities (Chankseliani & McCowan, 2021; Sørensen et al., 2019). Kaldeway (2018) has pointed out that global 'grand challenges' have started to dominate scientific discussions and policymakers' funding schemes in the twenty-first century. The received funds both transform and raise societal expectations towards HEIs' contributions to CC. Political steering is also manifested in the fact that many HEIs have adopted strategies around the United Nations Sustainable Development Goals (SDGs) and grand challenges like CC (Chankseliani & McCowan, 2021).

Between politics and HEIs lies the realm of higher education policy (cf. Capano & Jarvis, 2020), which forms the backbone for political steering and resource allocations to HEIs. Following New Public Management-inspired reforms (mid-1990s onwards), centred on quality, performance and accountability (Pinheiro et al., 2019), HE steering policies, in most countries, are based mainly on numerical targets on enrolments, degrees and publications. There is a strong cultural and legislative presupposition almost everywhere that academic inquiries are and should continue to be free of external influences. Hence, there is no guarantee that research will orientate itself towards CC. That said, the steering and orientation of research, by national funding agencies and other major donors as well as HEIs' central administrations, is done more indirectly; handled via institutional strategies (which tend to be rather symbolic and political in their nature) and the allocation of research funds. Increasing competition for scarce resources implies that such efforts are difficult to ignore by most academics, who require funding to undertake their desired inquiries regardless of their long-term impacts in society. In other words, co-optation mechanisms (Selznick, 1984) are at play in science, and substantial differences exist between 'haves' and 'have nots' (Kwiek, 2018).

The outcomes of research are often risky and open to multiple future avenues. Moreover, different disciplines may, in principle, be able to contribute to CC mitigation in different ways (Johnsen et al., 2015). Not all universities are, for example, studying energy solutions, and there is a growing divide between STEM and non-STEM fields, including in the attention paid by policy audiences (Benneworth et al., 2016; Reale et al., 2018).

Aside from this relatively loose overall steering, a new tighter policy discourse is rising. A recent survey of 400 global academic leaders, researchers and heads of funding bodies found that 68% agree that the inability to communicate the benefits of research for wider society could become an existential risk to the academic enterprise (Dixon, 2023). This problem was connected to the lack of common frameworks or methodologies (56% of the respondents), absence of consensus on what constitutes impact (48%), scarce resources for change (45%) and resistance from institutions and researchers alike (27%).

A major problem is that research impact tends to be slow and incremental. In an ongoing study by the book editors together with another colleague, we have analysed the UK's Research Excellence Framework (REF, 2021) 2021 database by using the search word Climate Change. The search produced 637 hits. The individual studies demonstrated good work and incremental improvements (e.g. improved fuel economy for airplane turbines, improved legislation, digital solutions, social and technical improvements and innovations, the latter in areas like engineering and physics, use of bioenergy, improved use of electricity in transport, etc.), but no clear signs of system-level solutions, game changers or major breakthroughs. These would require huge resources, and major cooperation and time. Studies show that innovation and its multifaceted societal effects do not unfold on a single event but instead requires different stages of development and cooperation amongst different parties (cf. Hage & Meeus, 2009). Penicillin, electricity and Apple's Macintosh all were required decades before they could be utilised on a large scale and have a quantifiable economic impact (Satell, 2017). Implementation requires industrial cooperation and applications.

Companies run the energy business and are crucial in investing and using green energy solutions, which cannot be put into practice without utilisation by these same companies, nor without the tacit support by investors funding risky enterprises. For example, erecting windmills is mainly a task for engineering and legal companies, rather than researchers per se. There are already many proven science-based solutions available to aid green transitions. The scientific basis of these solutions has been, for the most part, well-established for decades or even more than a century in some cases; in other words, they are 'known knowns'. But once investments in factories and other major facilities have been made, the primary intention within a globally competitive capitalist system is to run these facilities for decades, regardless of the negative impacts their operations may pose to the environment and local resources. If today's investments are not based on green energy solutions, due to a lack of financial incentives and/or political will (regulatory requirements), this may slow down any changes towards cleaner technologies, hampering the ambitious green transition targets set by bodies such as the EU.

As an example, a key player in the energy sector, the oil and gas industries, invested only 2.5% of its total capital spending into clean energies in 2022, which accounts for only 1% of total clean energy investments globally (IEA, 2023). The latest available figures on global coal power capacity seem to suggest an upward trend (Guardian, 2024a). Achieving the target scenario of a 50% investment rate would require close cooperation between governments, companies, shareholders and investors, as stated in a recent report by the International Energy Agency (IEA, 2023).

All in all, companies concentrate on their businesses, given that they are motivated by the market economy's competition and legislation (their main or sole task is to secure profits for shareholders). Their timeframe for reporting is rather short, as a quarterly framework is often in place. Here, revenue is the focus, and, unfortunately, fossil fuels currently provide the cheapest energy source on many occasions. In some larger companies there is room for R&D, but the incentives for change would need to come from the political sphere, and they often tend to be lacking.

In addition, *citizens'* consumption and behaviour patterns are a big part of the overall global emissions buildup. Populations across the globe require cheap energy for warming, cooling, cooking and transport, among other things. The public at large could make a strong political force. Protests for CC have existed for some time, most notably headed by the Swedish personality Grete Thunberg on behalf of future generations. Climate inaction was recently found to be a breach of human rights (Guardian, 2024b). However, the system providing energy—with its dynamics and networks of power—may be too complex for many laymen, compared to the fact that the current one works in everyday life. Simon Stiell, the current UN climate chief, has recently warned that climate finance by major supranational players like the World Bank must embark on a 'quantum leap' or risk facing 'climate-driven economic catastrophe', particularly to the developing world (Guardian, 2024c).

Moreover, it has been estimated that demand for energy is increasing as the world's population is still growing, which in turn intensifies CC. Currently, the world's poorest 4 billion people consume 5% of the total energy enjoyed by those living in developed countries. If the lowest consumption section rises to 15%, global energy consumption would increase by the equivalent of an additional United States' worth of demand (World Nuclear Association, 2022).

In short, all the key stakeholders play a major role in tackling CC in the fundamental energy sector. In principle, they all share the overall goal of CC mitigation, but their motivations differ, and they are trapped in different realities, have their own inertias and react to different time spans, obstacles and drivers. The timeframe, aims and 'rules of the game' differ, as shown in Table 1.1.

Becher (1989) has noted that, for an academic reputation is the same as power and money are for a politician and a businessman, respectively. Effective cooperation between the system-wide players to tackle CC would be needed, and would be a way forward, but that is often imperfect. As a result, the problem of CC tends to circulate between the stakeholders. At present the nexus politics—HEIs—corporations—citizens—as a global system seems incapable of substantially reducing emissions. On the contrary, global emissions have constantly increased since the 1800s, and many contend that we are heading in the wrong direction, although many countries have indeed made progress (The World Counts, 2023).

Supranational efforts in COP 27 (2023) to agree on emission cuts and strategic responses have taken place but fallen short, at least in the light of increasing global emissions. COP 28 (2024) added a statement that humanity should get rid of fossil fuels and aim at tripling global renewable power capacity by 2030. With such targets, a key issue is time. In the 2016 Paris Agreement, the EU committed to reduce its greenhouse gas emissions by at least 40% by 2030 compared to the 1990 level. In 2021, the goal was tightened to at least a 55% reduction by 2030, reaching climate neutrality by 2050.

Table 1.1 Key enablers: green transitions

	State (politicians)	*HEIs (academics)*	*Markets (companies)*	*Community (citizens and consumers)*
Key intention	Power and influence	Knowledge and reputation	Profits and revenue	Quality of life and self-realisation
Time span	Short (period of rule)	Long (depending on key persons and funding)	Short (quartile → strategy)	Generational dependent—from short to long term
Major assets in relation to CC mitigation	Right to make laws and to regulate; to allocate public funds	Possibility to study, develop and innovate based on research	Possibility to invest and implement technical solutions. R & D	Exercise democratic power and consumer freedom
Tends to avoid	Unpopular decisions, losing competitive position	Risk of losing reputation, funding and competitive position	Risk of losing money or competitive position.	Short-term (individual) sacrifices for long-term (community) benefits
Key external influences	Citizens voting behaviour	Political steering and funding	Demand from markets, competition	Widespread cultural norms and lifestyles (social media zeitgeist)

Source: Authors' own

Figure 1.2 highlights the interrelations amongst key actors at the system level. The key aspect here is that all key players are needed for fruitful progress in CC mitigation. It should be noted that the figure demonstrates *only* social interrelations, but the problem of CC is also physical in a hard sense: real CC is caused by societies' excessive emissions, which cannot be absorbed by the earth's natural systems. Excessive CO_2 emissions (as a physical entity) will remain in the atmosphere for hundreds or thousands of years (Taalas, 2021). The problem *could* be mitigated if successful cooperation with all the parties was met, and emissions could be cut. Another reservation, as noted, is that the problem of CC is global, whereas political decision-making, and business goals and strategies are (in

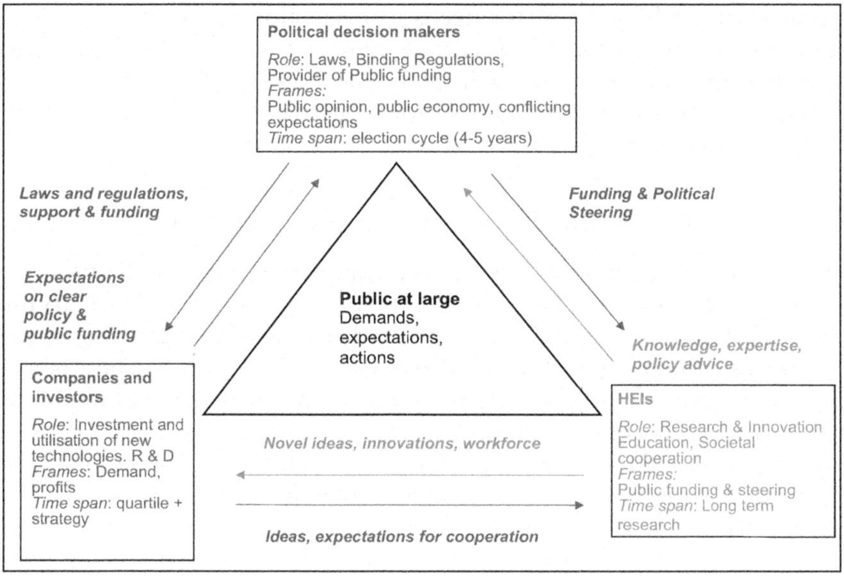

Fig. 1.2 Key players in CC mitigation: a systemic view

many cases) more localised, national or particular to the operator or area in question. The decisions are typically made in competition between other operators in the political or business sphere. There are ongoing regional (EU) and global (UN, e.g. SDGs) efforts to devise collaborative frameworks for CC mitigation, but verbal commitments often do not coincide with real efforts in the field, despite the good intentions of many key actors, HEIs included. For example, studies show that the "SDGs are found to be ***weak on agency***, with ***limited obligations*** to governments and none to business or consumers. They focus on state and impact, ***neglecting the pressures*** and supporting ***counterproductive drivers***" (Spangenberg, 2017, p. 311; emphasis added).

HEIs' basic tasks—research, education, innovation and engagement—still apply and form the basis of the work within the HE sector—albeit with increasing political pressures and 'strategic' promises to 'solve' the wicked problem of CC (Kekäle & Pinheiro, 2025). For the other key actors there are framing loose realities as well. Political decision-makers are strongly affected by public opinion (which is increasingly divisive and polarised); the business sector needs to make a profit to invest in green technologies. The demand for change counts much for these actors, and the time span for decision-making tends to be short: for politicians one election period, for the business sector a quartile—and for HEIs a much longer time-frame (Fig. 1.2).

Changing HE Policy and Governance

Above, we have simplified the highly complex phenomenon of CC by dealing with some key players or stakeholders especially in relation to the most fundamental energy sector. The fundamental role of greenhouse gas emissions and energy sector in CC was one of the logical starting points for this edited book. However, as a comparative basis the case studies of the book largely concentrate on empirical accounts of national HEI/CC policies and on HEIs' roles in CC mitigation. The basic idea is to map the territory for the first time, to maintain research freedom for each of the researchers involved with this volume and to avoid extremely complex comparisons between different influences and sectors. That being said, the editors provide authors with a guiding analytical framework which is revisited in the conclusion chapter to make broader sense of the empirical findings across cases.

As stated, in this volume the empirical contributions concentrate mostly on the policy-HEIs nexus in CC mitigation. The nation state's governance of HE systems plays a critically important role in shaping the domestic academic landscape in the quest to foster access, equity and quality across teaching, research and service/innovation. State-level governance mechanisms provide the necessary legal and technical frameworks for the effective functioning of HEIs, guiding decision-making, resource allocations and academic policies (Amaral et al., 2002).

State governance is crucial in ensuring that HE remains accessible and equitable, regardless of individuals' socioeconomic backgrounds or geographical location, and in maintaining high standards of quality and academic excellence. Regulatory bodies, such as accreditation and funding agencies, play a vital role in reviewing and monitoring HEIs to uphold proper academic standards and adequate means of conduct, such as the protection of institutional autonomy and academic freedom (Smeby & Stensaker, 1999). State governance can also facilitate the establishment of mechanisms for programme evaluation and curriculum development, promoting relevance and responsiveness to societal needs

(Mok & Yue, 2015). Through periodic assessments, the state strives to ensure that HEIs are constantly striving for improvements and are able and willing to meet the shifting demands of local, national and global labour markets and society more generally (Beerkens, 2020). The recent orientation towards SDGs is to a large part a result of political steering. As the superstructure (Clark, 1983), state governance plays a critical role in public resource allocations and in mediating strategic planning processes within HEIs. The state's ability to allocate funds in an equitable and efficient manner provides the required operational framework for HEIs to operate effectively, responsibly and in an accountable fashion.

The governance of HE systems by the state is interconnected with social development and economic growth at the local, regional and national levels (Benneworth, 2018; Schofer et al., 2000). By aligning sector-wide policies with national development goals, the state can address skills gaps, stimulate innovation and contribute to economic competitiveness. In so doing, it can enable the formation of research clusters, partnerships with industries and entrepreneurial activities, fostering a knowledge-based economy (Cai & Liu, 2015). At a broader level, state governance nurtures the integration of social responsibility into the fabric of HE systems and institutions (Sørensen et al., 2019). By promoting community engagement, social entrepreneurship and civic participation, state-guided frameworks enable students and faculty to address pressing social issues (like CC) and contribute to nation-building (Johnsen et al., 2015; Watson et al., 2011).

Higher education policy is a changing parameter which reflects the complex interplay between societal needs, economic forces, political imperatives and institutional aspirations. Rapid changes in policies easily contribute to slow responses by HEIs. Policy shifts tend to remain cascading: new emphases add to the kaleidoscope of expectations and influences, thus adding complexity within HEIs and at the system level (Clark, 2008; Kekäle & Pinheiro, 2024). To illustrate this, we shortly identify several policy shifts in the governance of (European) HE. The exclusive focus on Europe is a pragmatic one, and serves to illustrate ongoing reform and institutional dynamics, many of which are similar to other geographic contexts.

Starting in the 1990s, most governments across the world shifted from a policy of top-down state steering of HE towards that of regulating at a distance, by setting the rules and the allocation of resource pools (Jongbloed et al., 1999). In Europe, this led to a change in the nature of the social contract—from trust to accountability—between HE and society, brokered by the state (Neave & Van Vught, 1991). This, in turn, has enabled the prevalence of multiple stakeholders—and their strategic agendas—in the functions and daily affairs of HEIs (Pinheiro, 2015). One of the most notable policy shifts occurred in the 1990s with the adoption of the Bologna Process, aimed to create a single European Higher Education Area (EHEA) with comparable and compatible national HE systems (Witte, 2008). Bologna also encouraged a focus on quality assurance, student-centred and lifelong learning, all of which have contributed to a more integrated and diverse European HE system (Veiga, 2023). Another significant shift in European HE pertains to the increasing emphasis on employability and entrepreneurship. Policymakers across Europe recognised the importance of developing a skilled and adaptable workforce in a globalised economy and have responded by promoting initiatives such as work-integrated learning, entrepreneurship education and graduate employability (Römgens et al., 2020).

A third major policy shift in European HE reflects the growing importance of digitalisation and innovation. Technology has the potential to revolutionise the entire HE sector in terms of accessibility, affordability and quality. HEIs have responded by developing strategies around e-learning, open educational resources, massive open online courses (MOOCs) and data analytics, among others (Pinheiro et al., 2023a, 2023b). A fourth significant policy shift relates to the increasing focus on internationalisation, particularly across the European continent. Despite, or because of, the rise of neo-nationalistic tendencies across societies and politics (cf. Douglass, 2021), EU policymakers have re-invigorated the importance of preparing students for a globally interconnected world (and increasingly integrated EU region), and have, as a result, encouraged the development of international partnerships (strategic alliances), student and staff mobility, and intercultural competencies (Pinheiro et al., 2023c). The EU's 'flagship' *Horizon Europe* programme, for example, supports international cooperation in research and innovation, promoting interdisciplinarity and collaboration across borders. This shift has contributed to expanding academic networks, enhancing diversity and promoting cultural exchange (European Commission, 2024), alongside fostering resilience across scientific institutions and critical sectors like health and wellbeing and security (European Commission, 2023a). However, it also poses new challenges to HEIs, such as ensuring the quality of international programmes, mitigating brain drain and respecting cultural and linguistic diversity. More recently, Horizon Europe is seen as instrumental in stimulating the development of knowledge and technologies deemed crucial for tackling global 'wicked' challenges such as CC and the transition to a more sustainable use of the earth's limited resources (European Commission, 2021, 2023b).

Finally, the COVID-19 pandemic has accelerated policy shifts in European HE, prompting policymakers and HEIs alike to adapt to new realities and challenges. The pandemic has, amongst other aspects, reinforced the importance of digital learning, student support and mental health services (Pinheiro et al., 2023b). It has also highlighted the need for flexibility and resilience (Young et al., 2024; Young & Pinheiro, 2022), as providers had to rapidly adapt to new circumstances while retaining function, by, for example, shifting to online learning and remote work (cf. Laterza et al., 2023). COVID-19 has also underscored longstanding issues such as equity, access and affordability, prompting policymakers and HEIs across Europe and beyond to reconsider traditional models of delivery and seek new domestic and international partnerships and collaborations (Pinheiro et al., 2023c).

This discussion on policy shifts demonstrates that political steering tends to have multiple and changing focus areas while political issues such as equity, access and affordability tend to remain important. New additions, such as the mitigation of CC, end up adding to the complex whole of diversified, multilayered landscapes of HE policies and governance. In any event, policies are an essential tool in governing HE systems and institutions as they provide a framework for decision-making and strategic resource allocations, help establish operational and ethical guidelines (Pinheiro et al., 2019). The policy mix available to governments is multifaceted, covering a broad range of mechanisms, tools and issues (Musselin & Teixeira, 2014). Given the complexity inherent to both HE systems and institutions, governments tend to adopt hybrid steering approaches that resort to a mix between traditional top-down (government-centered) and market-based approaches (Gornitzka & Maassen, 2000). Salient issues across most national HE systems include, but are not limited to, aspects like academic standards, financial management, student affairs, teaching quality, curriculum innovation, diversity and inclusion, accountability, as well as graduate employability (cf. Hazelkorn et al., 2018).

Effective policies also contribute to addressing the critical topic of accountability (Stensaker & Harvey, 2011). By establishing procedures for internal decision-making processes (governance) and conflict management, government policies seek to ensure transparency and fairness.

Furthermore, policies help ensure that HEIs operate within legal and ethical standards, providing a suitable institutional framework for them to comply with government regulations and uphold ethical principles. Notwithstanding, tensions exist between government efforts to foster control and accountability and the required trust necessary to execute institutional and academic tasks that safeguard cherished norms, values and traditions (Hansen et al., 2019).

Comparative Public Policy: A brief overview of key approaches

There is a wide array of policy instruments used in public policy research to categorise, analyse and compare approaches across sectors and national settings (*cf.* Dodds, 2018). One widely recognised typology is Vedung's (1998) trichotomy of *carrots*, *sticks* and *sermons*, representing different means through which policymakers can induce desired behaviour amongst local actors, HEIs included (see Capano & Jarvis, 2020; Klenk et al., 2022), in the context of reform implementation and/or change. *Carrots*, refers to the means (often economic ones) used as primary incentives to encourage certain courses of actions. This may involve providing additional resources or benefits to policy recipients. For example, a government may offer financial grants or rewards to HEIs that engage in specific behaviours related to the desired outcome (e.g. mergers or amalgamations, see Pinheiro et al., 2016). *Sticks*, represents regulatory measures aimed at restricting or punishing certain types of behaviours, such as lack of compliance with the prescribed policies (see Gornitzka et al., 2005). Examples include laws, directives and decrees that force compliance with specific actions or behaviours. Finally, as policy instruments, *sermons* centre on the use of information and targeted or strategic communications to influence actors' behaviours through framing and messaging. This can involve shaping perceptions and values related to the desired policy outcome (see Bordin & Pinheiro, 2022; Chou et al., 2023). Public awareness campaigns, educational programmes and information dissemination are all typical tools under this policy instrument.

It is important to highlight that the aforementioned (3) categories are not mutually exclusive per se and can often be combined in creative ways (Klenk et al., 2022), adding to the complexity of the policy process (Trondal et al., 2022). Different policy instruments can be employed depending on the specific context, policy goals and stage of the policy process. The concept of *policy mix* (Cocos & Lepori, 2020) proves useful in understanding the complex and dynamic intricacies (and contradictions) underpinning policy implementation as a dynamic and non-linear process (Gornitzka et al., 2005). Moreover, it reinforces the notion that policy instruments are the fundamental building blocks of governance both within and across sectors of the economy and polity (Capano et al., 2015).

Capano and Jarvi's (2020) work on the impact of salient exogenous dimensions, such as globalisation and Europeanisation, on the governance of HE systems provides an additional analytical framework for comparing developments across national settings. The authors refer to four relatively distinct key governance mechanisms.

The first is *convergence*, a process driven by global and European policy agendas seeking to create unified and harmonised HE and research areas (see Maassen & Olsen, 2007). The adoption of similar policy frameworks, such as the intergovernmental Bologna Process (Sin et al., 2016) and, more recently, the supranational-led European University Initiative (Gänzle & Pinheiro, 2025, in press), amongst others, has resulted in similar governance structures and mechanisms across European HE systems and institutions.

The second mechanism pertains to patterns of *divergence* or isomorphism. Despite the observed convergence in the adoption of similar policy frameworks or mixes, studies also show significant national differences as regards the implementation of governance reforms in the light of historical contexts (path dependencies) alongside local norms, values and (institutionalised) traditions (Musselin, 2009; Veiga, 2023). The latter shape the ways in which the adopted policies are interpreted and implemented by local actors, resulting in variations in the actual practices and outcomes of governance reforms.

The third mechanism identified by Capano and Jarvis (2020) is that of increasing *hybridisation* of governance models, centred on the coevolution and coexistence between national and supranational dimensions. In this context it is observed that, while European-level policies exert influence on national governance systems, the former often merge with or become embedded in existing national traditions and practices. This results in complex, multi-level governance arrangements that involve a combination of supranational, national and institutional actors (cf. Chou et al., 2017).

As for the fourth and final mechanism, Capano and Jarvis (2020) shed light on the effects of governance on both the *autonomy* and *performance* of HEIs and the national systems in which they are deeply embedded in (see Pekkola et al., 2021). It is observed that convergence patterns may result in a decline of institutional autonomy (as well as trust), as governmental policies become more standardised and centralised (cf. Hansen et al., 2019). Strategic metrics tend to be part and parcel of assessment procedures within teaching and/or research (e.g. Kekäle, 2008), which often results in *gaming* as a widespread (strategic) behavioural pattern amongst local actors (Aboubichr & Conway, 2023). Recent studies from the Nordics suggest that policy shifts geared towards stronger collaborations both within and across HEIs are tightly intertwined with rising global competition for talent, funding and prestige (Pinheiro & Pulkkinen, 2023). That said, there are plenty of circumstances where the effective implementation of HE reforms results, in the mid- to long run, in greater accountability, performance and quality assurance (Gornitzka et al., 2005; Hazelkorn et al., 2018).

In short, Capano and Jarvi's insightful work highlights the tensions between convergence and diversity in the governance of HE systems and institutions. While Europeanisation and globalisation contribute to greater convergence in policy frameworks, the actual implementation and outcomes of these reforms vary (sometimes greatly) across national settings. Hybrid governance models emerge as a result, combining national traditions with supranational influences. Finally, the observed effects of governance reforms on institutional autonomy and performance are complex and multifaceted, and laden with critical trade-offs between standardisation and accountability.

Seen from an instrumental (means-ends) perspective, policy instruments are ultimately meant to have a tangible effect (either change or continuity) on a specific policy sub-field (Christensen et al., 2019), like HE. When governments adopt policy mechanisms like performance-based funding or changes in curriculum structures, the expectation is that such measures will, in the mid- to long run, result in enhanced quality across the board. Nonetheless, it is important to highlight that policy instruments are *equifinal* in nature (Capano, 2024). This basically means that the same policy outcome can be achieved through different types of instruments, depending on the local contexts and the existing policy legacy. This is especially the case when considering the micro-design of policy instruments since their intrinsic characteristics and dimensions might have a significant impact on their overall effectiveness. Thus, the effectiveness of policy instruments largely depends on a wide range of factors, namely, the policy context, the specific characteristics of the instrument and the pre-existing policy landscape (Capano, 2024; Capano & Jarvis, 2020).

Higher Education and the Wicked Issue of Climate Change

Given the relatively challenging policy backdrop sketched out above, the aim of this edited volume is to illuminate the role that policy plays in terms of providing an enabling framework for HEIs to tackle CC: directly, through their core missions of teaching, research, innovation and engagement; and, indirectly, through their role as bastions and curators of world knowledge, values and cultures. The volume aims to describe the overall situation across a wide variety of selected countries and three world regions: Europe, Asia and South America. These cases are by no means exhaustive, nor do they claim to be representative of the wider picture across the globe, but they do represent a relevant micro-cosmos from which larger lessons can be derived. Each national HE system is embedded in their own specific political, economic and legislative contexts. The writers were given considerable leeway to provide comparative basic information on the matter.

The volume seeks to highlight the *drivers*, *dynamics* and *effects* of the complex policy processes underpinning the role of HEIs and knowledge more generally. While many fundaments in the mitigation of CC may not be in HEIs' direct control, there are still many things that can be done by institutions and academic groups across a wide spectrum of disciplinary, sub-disciplinary and inter/trans-disciplinary activities.

Higher education institutions have been recognised as having a critical role in the fight against CC. As is the case with most public sector organisations, universities, particularly larger ones, have significant greenhouse gas emissions due to energy consumption in their buildings, transportation and other related activities. Many universities have started installing solar panels, adopting energy-efficient habits and investing in green buildings (Fissi et al., 2021). Through their primary activities, HEIs the world over have started addressing sustainability issues across the board. Chief among them include adaptations to existing and the development of innovative curriculum structures that address CC and its impact on society and ecosystems (Leal Filho et al., 2019; McCowan et al., 2021). Many are establishing dedicated sustainability programmes at the undergraduate and graduate levels to provide interdisciplinary training necessary to tackle CC in a comprehensive and holistic way (McCowan et al., 2021). In the realm of knowledge production, some institutions have established specialised research units or institutes to facilitate cross-disciplinary collaboration and inquiry on climate-related issues (Cristofoletti & Pinheiro, 2023). They have also engaged with a wide variety of stakeholder groups to co-create knowledge that addresses climate change, alongside other sustainable development goals (Agusdinata, 2022; Alcántara-Rubio et al., 2022). On the governance front, students and faculty are encouraging their universities' leadership structures to establish sustainability committees or green energy clubs to promote renewable energy initiatives (McCullough et al., 2018; Spira et al., 2013).

According to Witte (2021, 93) science and research make three major contributions towards tackling the climate crisis, namely:

1. Measuring and understanding CC: providing the empirical basis for policy decisions and political action (according to our estimation, about 40% of the 637 REF case studies we analysed fall into that category, Kekäle et al. in this volume).
2. Developing climate-friendly technologies and economic and social models that allow humankind to limit global CC to bearable levels (mitigation) (roughly 30% of the REF cases belonged into that category). Here, energy usage would be among the key issues.
3. Devising coping and adjustment strategies, helping humankind to deal with CC, both current and future (adaptation). (about 30% of the REF cases belonged into that category).

Arguably, HEIs best serve society by bringing vital information to decision-makers, but increasingly also in providing solutions for new ways forward and for solving bottlenecks around the green (energy) transition, to help developing coping and adjustment strategies and in educating students and citizens at large. As a part of the national mitigation targets, HEIs need to do their part (typically in real estate management, or diminishing carbon footprint or air travel), but going beyond that would require intensified focus of resources and strategic efforts. Stensaker and Hermansen (2023) have recently noted that HEIs either focus their resources and capabilities on pragmatic efforts at their disposal such as educational offerings, research profiles or academic strengths (for serving the strategic goal of CC mitigation), or, instead, strategically engage in symbolic, greenwashing activities with little effect on institutional life.

Yet, despite their good intentions, as rational actors, HEIs and their leaders respond to the signals and incentives provided by policymakers and the institutional (rules and regulations) frameworks that both enable and constrain their strategies. Higher education policy clearly operates in the social and political realm that can deviate from the reality of the actual physical emissions produced by burning fossil fuels. Still, policies represent the official system level's attempt to do something about the problem. In any given national political system, the core policy issues of relevance to society often return to HEIs by using funding and regulations as steering mechanisms; both with 'political push' and 'institutional pull' for funding. Hence, these efforts try to move HEIs from relatively autonomous fiduciary institutions into instruments for reaching pre-determined policy goals and shifting ideological agendas (Maassen & Olsen, 2007). This transformation of HEIs into 'policy agents' risks destroying the legitimacy and autonomy of HEIs and science that lies at the heart of their raison d'être as public goods (Marginson, 2016). Arguably, the view shared by the authors of this volume is that HEIs *can* and *should* play an active role in tackling CC, however they should do so in the light of their wider public mandates and whilst safeguarding their own institutional integrity alongside that of the multiple knowledge domains and communities (scholarly and of practice) that are hosted within it.

Description of the Contributions

Chapter 2, by Kekäle and colleagues, illustrates two contrasting national policy approaches to deal with the role of science and HE in CC mitigation in Europe. Using the Dutch CoARA and the UK's REF as illustrative cases, their analysis found different drivers, origins and backgrounds for HE policy and steering, alongside different modes of the 'social pact' between HE and society in relation to CC. The findings show that in the Dutch case HEIs are expected, and largely subscribe, to a more central role in CC mitigation, whereas in the UK this is done in a more indirect and less substantive manner, by resorting to intermediary bodies and a marginal level of institutional commitment. In so doing, the chapter sheds light on the importance of national traditions and institutional conditions and regulatory aspects in shaping HEIs' responses to CC.

In Chap. 3, Gibson and Hazelkorn, set out the nature of Ireland's policy response to the climate emergency and the role of the HE sector in it, across the various policy realms (national, sectoral and institutional). In so doing, the authors question what these elements amount to in practice. Adopting a critical policy approach, the chapter interrogates whether relevant policies merely reframe existing concerns, react to external pressures or amount to proactive responses appropriate to the enormity of global warming. The first part of the chapter outlines government policies regarding climate change generally, and some of the unusual features of the Irish policy context, which give considerable weight to the voice of civil society. From this, the authors move to consider HE policy set out by government, before addressing the institutional policy and academic research in relations to CC. The chapter concludes with an argument for the specific role for HE, 'thinking in an emergency', and a call demanding transformational change.

Chapter 4, by Reale and Spinello, sheds light on the structure and the effects of the Italian government's R&D policies for sustaining or increasing academic research on topics related to CC. The authors apply a conceptual framework based on the 'opportunity approach' to learn more about the effectiveness of potential incentives for CC research. Using multiple data points, their analysis shows that—despite a profusion of policy documents highlighting the important role that science plays in tackling CC—over the last few years Italian research policies have offered very limited funding opportunities to HEIs for CC research. As such, the funding has primarily supported short-term projects which fall short of effectively addressing the complex challenges facing society. To address this, it is argued, a more structured and comprehensive approach is required, with long-term financial resource mobilisation supplemented by follow-on funding.

David Charles in Chap. 5 provides an overview of how UK universities have committed to the CC objective, including their role as engaged or civic universities with a responsibility to take a leadership role within their communities. Amongst other aspects, it is outlined that the successful implementation of actions to reduce CC gas emissions, and to assist the wider community of firms and institutions in mitigation and adaptation also depends on their partners, and especially the support from central government. The data shows that whilst UK universities have been active in the CC agenda for several decades in some respects, their involvement has not always been consistent across all the fields of activity which could be undertaken. The greatest level of activity has been in the research field where the UK government has provided considerable funding for CC related research. That said, domestic universities have played an important international role in the development of knowledge. The overall picture is rather positive, and many universities have declared a climate emergency, so showing their commitment, but implementation has depended on the direction and support given by national government: strong on research, and weak on regional partnership development.

Moving to Central and Eastern Europe (CEE), Chap. 6 by Antonowicz and colleagues maps out the efforts undertaken by Polish HEIs to tackle

CC. The chapter examines each of the three HE levels (macro, meso and micro) separately as they are, to a large degree, autonomous and driven by different rules and actors. Overall, the study presents a rather gloomy picture of the Polish HE system, being politically and institutionally unengaged in addressing one of the biggest crises facing humanity. The analysis shows that the government's interest in tackling CC has been rather modest and limited to the political commitments stemming from Poland's EU membership. The policy initiatives and financial resources focus mostly on private businesses as vehicles of transformation for the energy sector. As public institutions, HEIs seem to be (still) rather immune to growing external calls for stronger engagement in tackling CC. Instead, the study shows that actors within HEIs pay considerable attention to reputational games driven by international global rankings. That said, the study does provide some hope for the future in the ample and continuously increasing interest of individual academics and research teams in pursuing problems related to CC mitigation.

Remaining in the CEE region, Chap. 7 by Kohoutek presents an exploratory enquiry into the extent to which CC issues figure as an agenda in HE policies in the Czech Republic (CR). The chapter sets the enquiry into a broader framework of HEIs' climate protection (in)actions including the relevant international implications. The findings show that the European Green Deal clearly served as a driver initiating the green agenda for the Czech HE sector, in the form of the CR's Recovery and Resilience Plan (RRP). The RRP was found to show little support of HE-related climate-saving activities per se, with the focus put on ancillary activities represented by phasing-in new study courses or programmes, curricular transformation, ICT technologies or biomedical research. The author concludes that, for now, CC is not a priority for Czech HE policymaking system-wise, with the initiative so far in the hands of voluntary, pro-active actions by a handful of individual HEIs.

Moving to Asia, Chap. 8 by Yu, Linlin and Botao show how, in recent decades, the Chinese government has undertaken a range of policies, measures and initiatives to tackle CC. This study is based on a literature review and a systematic examination of pertinent documents. It shows that the national government has actively been engaging in global climate governance and yielding favourable outcomes. Universities are widely recognised as having crucial responsibilities in both adapting to and mitigating CC. Some universities have introduced master's and doctoral programmes focused on sustainable development to educate professionals in this field.

They have also established research institutions dedicated to sustainable development and collaborated with global cooperation platforms. In addition, they have implemented several measures to support both domestic and global policies in addressing CC. These initiatives include focusing on scientific research, talent development and international interactions relevant to climate change challenges. The chapter analyses the development process, policy backdrop, practical activities and outcomes of Chinese universities' reactions to CC, as well as identify future paths.

In South America, Axel-Berg and colleagues (Chap. 9) investigate the different roles taken by elite traditional research-intensive institutions in southeast Brazil and research organisations in the Amazon region. As the result of a historical imbalance of funding and capacity building, HEIs in the Amazon, a region with some of the highest demands for science on earth, have largelly been underfunded and underdeveloped. Their analysis shows that, while this scenario is rapidly changing, intra-national collaboration has proven to be a vital mechanism for sharing expertise and facilities that these (more remote) institutions would otherwise struggle to access. Institutions in the Amazon region have witnessed a huge increase in bioeconomy-related research in the last five to ten years, and now constitute a significant amount of the knowledge produced on sustainable development of the biome. Alone, however, these institutions do not possess the resources to satisfy the demands for scientific knowledge needed to create a sustainable bioeconomy. The findings highlight the importance of joint funding initiatives at the state level whilst simultaneously shedding light on the limitations of these initiatives, framed as they are by the usual approach adopted by public research funding in Brazil.

Chapter 10, by Felder, revisits a burgeoning topic in global HE in the scope of strategic alliances involving HEIs driven by purposes of internationalisation and competition. As a supranational body, the European Union has recently turned to HEI alliances as problem-solving tools for societal challenges. Since the combat against CC is assuming a central role in political agendas at the (sub)national and supranational level, it is expected to also shape the logics of HEIs' alliance formation. This contribution asks how HEI alliances are affected by the impetus for a stronger contribution by HEIs to tackle CC. The analysis investigates the role of CC for the missions and cooperation results of HEI alliances in the European Education Area. In terms of missions, the author finds that HEI alliances cooperate in climate issues due to the profiles of the involved HEIs and driven by societal demands and funding stipulations. Regarding cooperation results, the analysis shows that alliances primarily incorporate

CC action into short-term joint teaching formats and into the exchange of researchers with shared interests. Moreover, alliances serve to exchange practices to make campuses and different aspects of university life such as student mobility more sustainable. The chapter concludes that choosing CC mitigation as an alliance mission serves strategic purposes such as unifying its members, securing funding and/or strengthening one's competitive or market positioning both nationally and globally.

Finally, in Chap. 11 the editors review key themes and trends across the empirical cases at the heart of the volume and revisit the conceptual stakeholder model presented in Chap. 1, paying tribute to the complexity inherent to the problem of CC mitigation. The volume concludes with suggestions for future studies in this important and increasingly contested domain of social, economic, cultural and political life, one in which HEIs should not and cannot be left behind.

References

Aboubichr, B., & Conway, N. (2023). The gaming of performance management systems in British universities. *Human Relations, 76*(4), 602–628.

Agusdinata, D. B. (2022). The role of universities in SDGs solution co-creation and implementation: A human-centered design and shared-action learning process. *Sustainability Science, 17*(4), 1589–1604.

Alcántara-Rubio, L., Valderrama-Hernández, R., Solís-Espallargas, C., & Ruiz-Morales, J. (2022). The implementation of the SDGs in universities: A systematic review. *Environmental Education Research, 28*(11), 1585–1615.

Amaral, A., Jones, G. A., & Karseth, B. (2002). *Governing higher education: National perspectives on institutional governance*. Kluwer Academic Publishers.

Becher, T. (1989). *Academic tribes and territories. Intellectual enquiry and the cultures of disciplines*. The Society for Research into Higher Education & Open University Press.

Beerkens, M. (2020). Evidence-based policy and higher education quality assurance: Progress, pitfalls and promise. In T. Leiber (Ed.), *Impact evaluation of quality management in higher education* (pp. 38–53). Routledge.

Benneworth, P. (2018). *Universities and regional economic development: Engaging with the periphery*. Taylor & Francis.

Benneworth, P., Hazelkorn, E., & Gulbrandsen, M. (2016). *The impacts and future of arts and humanities research*. Palgrave.

Bordin, E., & Pinheiro, R. (2022). The race to institutional change: The slow road to policy change in sexuality education. In J. Trondal, R. Keast, D. Noble, & R. Pinheiro (Eds.), *Governing complexity in times of turbulence* (pp. 146–168). Edward Elgar Publishing.

Cai, Y., & Liu, C. (2015). The roles of universities in fostering knowledge-intensive clusters in Chinese regional innovation systems. *Science and Public Policy, 42*(1), 15–29. https://doi.org/10.1093/scipol/scu018

Capano, G. (2024). Policy instruments. In *Encyclopedia of public policy* (pp. 1–6). Springer.

Capano, G., Howlett, M., & Ramesh, M. (2015). Bringing governments back in: Governance and governing in comparative policy analysis. *Journal of Comparative Policy Analysis: Research and Practice, 17*(4), 311–321.

Capano, G., & Jarvis, D. S. (2020). *Convergence and diversity in the governance of higher education: Comparative perspectives.* Cambridge University Press.

Chankseliani, M., & McCowan, T. (2021). Higher education and the sustainable development goals. *Higher Education, 81*(1), 1–8.

Chou, M. H., Elken, M., & Jungblut, J. (2023). Policy framing in higher education in Western Europe: (Some) Uses and (many) promises. In J. Jungblut, M. Maltais, E. C. Ness, & D. Rexe (Eds.), *Comparative higher education politics: Policymaking in North America and Western Europe* (pp. 231–257). Springer.

Chou, M. H., Jungblut, J., Ravinet, P., & Vukasovic, M. (2017). Higher education governance and policy: An introduction to multi-issue, multi-level and multi-actor dynamics. *Policy and Society, 36*(1), 1–15.

Christensen, T., Lægreid, P., Roness, P. G., & Røvik, K. A. (2019). *Organization theory and the public sector: Instrument, culture and myth.* Taylor & Francis.

Chrobak, U. (2021). The world's forgotten greenhouse gas. *BBC*, June 4. Retrieved June 24, 2024, from https://www.bbc.com/future/article/20210603-nitrous-oxide-the-worlds-forgotten-greenhouse-gas

Clark, B. R. (1983). *The higher education system: Academic organization in cross-national perspective.* University of California Press.

Clark, B. R. (2008). The problem of complexity in modern higher education. In B. R. Clark (Ed.), *On higher education. Selected writings, 1956–2006.* Johns Hopkins University Press.

Cocos, M., & Lepori, B. (2020). What we know about research policy mix. *Science and Public Policy, 47*(2), 235–245.

Cook, J., Nuccitelli, D., Green, S.A, Richardson, M., Winkler, B., Painting, R., Way, R., Jacobs, P., & Skuce A. (2013) Quantifying the consensus on anthropogenic global warming in the scientific literature. *Environmental Research Letters 2*(8), 024024.

COP 27. (2023). Delivering for people and the planet. UN Climate Action. Retrieved July 1, 2024, from https://www.un.org/en/climatechange/cop27

COP 28. (2024). COP 28: What was achieved and what happens next? UN climate action. Retrieved July 1, 2024, from https://unfccc.int/cop28/5-key-takeaways

Cristofoletti, E. C., & Pinheiro, R. (2023). Greening the University? Assessing the impact of sustainability and SDGs in universities' values and strategies. In W. L. Filho, F. Frankenberger, & U. Tortato (Eds.), *Sustainability in practice: Addressing challenges and creating opportunities in Latin America* (pp. 111–126). Springer.

Dixon, E. (2023). 'Existential risk' to research from failure to demonstrate impact. Sector leaders quizzed in Elsevier survey back shift to more holistic methods of evaluating scholarship. Times Higher Education Newsletter, November 7. Research assessment 'fails to encourage impact' | Times Higher Education (THE). Retrieved November 24, 2023.

Dodds, A. (2018). *Comparative public policy*. Bloomsbury Publishing.

Douglass, J. A. (2021). *Neo-nationalism and universities: Populists, autocrats, and the future of higher education*. JHU Press.

European Commission. (2021). From focus areas to EU missions: The EC focus area policy under Horizon 2020 (2017–2020). Directorate-General for Research and Innovation, Publications Office of the European Union. Retrieved June 7, 2024, from https://data.europa.eu/doi/10.2777/40351

European Commission. (2023a). Evaluation study of the European framework programmes for research and innovation for a resilient Europe – Final report – Phase 1 – Executive summary. Retrieved June 7, 2024, from https://op.europa.eu/en/publication-detail/-/publication/d3c6abb4-012e-11ee-87ec-01aa75ed71a1/language-en

European Commission. (2023b). The transformative nature of the European framework programme for research and innovation: Analysis of its evolution between 2002–2023. Directorate-General for Research and Innovation Publications Office of the European Union. Retrieved June 7, 2024, from https://data.europa.eu/doi/10.2777/73000

European Commission. (2024). Horizon Europe. Retrieved June 7, 2024, from https://research-and-innovation.ec.europa.eu/funding/funding-opportunities/funding-programmes-and-open-calls/horizon-europe_en

Fissi, S., Romolini, A., Gori, E., & Contri, M. (2021). The path toward a sustainable green university: The case of the University of Florence. *Journal of Cleaner Production, 279*, 123655.

Gates, B. (2021). *How to avoid climate disaster*. Vintage Books.

Gänzle, S., & Pinheiro, R. (2025, in press). From the 'European university' to the Europeanisation of universities in Europe: E pluribus multum. In A. Frame & B. Curylo (Eds.), *The European Universities Initiative and the 'Euro-Internationalisation' of European Higher Education*. Routledge.

Global Offshore Wind Report. (2022). Retrieved March 27, 2023, from https://gwec.net/wp-content/uploads/2022/06/GWEC-Offshore-2022_update.pdf

Goodreads. (2024). *Jean-Claude Junkner quote*. https://www.goodreads.com/quotes/454948-we-all-know-what-to-do-but-we-don-t-know

Gornitzka, Å., Kogan, M., & Amaral, A. (2005). *Reform and change in higher education: analysing policy implementation*. Springer.

Gornitzka, Å., & Maassen, P. A. M. (2000). Hybrid steering approaches with respect to European higher education. *Higher Education Policy, 13*(3), 267–285.

Guardian. (2024a). World's coal power capacity rises despite climate warnings. Retrieved June 7, 2024, from https://www.theguardian.com/environment/2024/apr/11/worlds-coal-power-capacity-rises-despite-climate-warnings

Guardian. (2024b). Human rights violated by Swiss inaction on climate, ECHR rules in landmark case. Retrieved June 7, 2024, from https://www.theguardian.com/environment/2024/apr/09/human-rights-violated-inaction-climate-echr-rules-landmark-case

Guardian. (2024c). World Bank must take 'quantum leap' to tackle climate crisis, UN expert says. Retrieved June 7, 2024, from https://www.theguardian.com/environment/2024/apr/10/world-bank-must-take-quantum-leap-to-tackle-climate-crisis-un-expert-says

Hage, J., & Meeus, M. T. H. (2009). *Innovation, science, and institutional change.* Oxford University Press.

Hainsch, K., Löffler, K., Burandt, T., Auer, H., Crespo del Granado, P., Pisciella, P., & Zwickl-Bernhard, S. (2021). Energy Transition Scenarios: What policies, societal attitudes, and technology developments will realize the EU Green Deal? Retrieved June 7, 2024, from https://www.researchgate.net/publication/354896800_Energy_Transition_Scenarios_What_policies_societal_attitudes_and_technology_developments_will_realize_the_EU_Green_Deal

Hansen, H. F., Geschwind, L., Kivistö, J., Pekkola, E., Pinheiro, R., & Pulkkinen, K. (2019). Balancing accountability and trust: university reforms in the Nordic countries. *Higher Education, 78*(3), 557–573. https://doi.org/10.1007/s10734-019-0358-2

Hazelkorn, E., Coates, H., & McCormick, A. C. (2018). *Research handbook on quality, performance and accountability in higher education.* Edward Elgar Publishing.

Head, B. W. (2022). *Wicked Problems in Public Policy.* Springer.

IEA. (2023). The oil and gas industry in net zero transitions online. Retrieved December 1, 2023.

Johnsen, H. C. G., Torjesen, S., & Ennals, R. (2015). *Higher education in a sustainable society: A case for mutual competence building.* Springer.

Jongbloed, B. W. A., Maassen, P. A. M., & Neave, G. R. (1999). *From the eye of the storm: Higher education's changing institution.* Kluwer Academic Publishers.

Kaldeway, D. (2018). The grand challenges discourse: Transforming identity work in science and science policy. *Minerva, 56*(2), 161–182.

Kekäle, J. (2008). The negotiation process toward the new salary system in the Finnish University sector. *Higher Education Management and Policy, 20*(3), 97–117.

Kekäle, J., & Pinheiro, R. (2025). Individual leadership in higher education: Partial solutions, unintended consequences, and the search for an alternative. In M. B. Rao, A. Singh, & M. P. Rao (Eds.), *Worldviews and values in higher education: Institutional governance, leadership, and capacity building* (pp. 127–149). Emerald.

Klenk, T., Antonowicz, D., Geschwind, L., Pinheiro, R., & Pokorska, A. (2022). Taking women on boards: A comparative analysis of public policies in higher education. *Policy Reviews in Higher Education, 6*(2), 128–152.

Kwiek, M. (2018). *Changing European academics. A comparative study of social stratification, work patterns and research productivity*. Routledge.

Laterza, V., Asante, M. O., Tømte, C. E., & Pinheiro, R. (2023). Implementing digital transformations in higher education following COVID-19: A Norwegian case study. In R. Pinheiro, C. Tømte, L. Barman, L. Degn, & L. Geschwind (Eds.), *Digital transformations in Nordic higher education* (pp. 243–266). Springer.

Leal Filho, W., Shiel, C., Paço, A., & Mifsud, M. (2019). sustainable development goals and sustainability teaching at universities: Falling behind or getting ahead of the pack? *Journal of Cleaner Production, 232*, 285–294.

Lockwood, B., & Lockwood, M. (2022). How do right-wing populist parties influence climate and renewable energy policies? Evidence from OECD countries. *Global Environmental Politics, 22*(3), 12–37.

Maassen, P., & Olsen, J. P. (2007). *University dynamics and European integration*. Springer.

Marginson, S. (2016). Public/private in higher education: A synthesis of economic and political approaches. *Studies in Higher Education, 2*(43), 322–337.

Masterson, V. (2022). Wind and solar generated 10% of global electricity in 2021 – A world first. World Economic Forum. Retrieved March 27, 2023, from https://www.weforum.org/agenda/2022/04/wind-solar-electricity-global-energy/

McCowan, T., Leaf Wilho, W., & Brandli, L. (Eds.). (2021). *Universities facing climate change and sustainability*. Körber Stiftung.

McCullough, B. P., Kellison, T., & Wendling, E. (2018). Formation and function of a collegiate athletics sustainability committee. *Journal of Amateur Sport, 4*(1), 52–81.

Mok, J. K., & Yue, K. (2015). Promoting entrepreneurship and innovation in China: Transformations in University curriculum and research capacity. In S. Schwartzman, R. Pinheiro, & P. Pillay (Eds.), *Higher education in the BRICS countries: Investigating the pact between higher education and society* (pp. 439–462). Springer.

Musselin, C. (2009). The side effects of the Bologna process on national institutional settings: The case of France. In A. Amaral, G. Neave, C. Musselin, & P. Maassen (Eds.), *European integration and the governance of higher education and research* (pp. 181–205). Springer.

Musselin, C., & Teixeira, P. N. (Eds.). (2014). *Reforming higher education: Public policy design and implementation* (Vol. 41). Springer.

Neave, G., & van Vught, F. (Eds.). (1991). *Prometheus bound: The changing relationship between government and higher education in Western Europe*. Pergamon.

Pekkola, E., Pinheiro, R., Geschwind, L., Siekkinen, T., Carvalho, T., & Pulkkinen, K. (2021). Nested hybridity and value definition in public higher education. In J. Vakkuri & J.-E. Johanson (Eds.), *Hybrid governance, organisations and society: Value creation perspectives* (pp. 59–80). Routledge.

Pinheiro, R. (2015). The role of internal and external stakeholders. In S. Schwartzman, R. Pinheiro, & P. Pillay (Eds.), *Higher education in the BRICS countries: Investigating the pact between higher education and society* (pp. 43–57). Springer.

Pinheiro, R., & Pulkkinen, K. (2023). Collaborating for competition? *Scandinavian Journal of Public Administration, 27*(1), 53–73.

Pinheiro, R., Geschwind, L., & Aarrevaara, T. (Eds.). (2016). *Mergers in Higher Education: The experiences from Northern Europe*. Springer.

Pinheiro, R., Geschwind, L., Hansen, H. F., & Pulkkinen, K. (2019). *Reforms, organizational change and performance in higher education: A comparative account from the Nordic countries*. Palgrave.

Pinheiro, R., Gänzle, S., Klenk, T., & Trondal, J. (2023a). Unpacking strategic alliances in European higher education. *Tertiary Education and Management, 29*(3), 213–228.

Pinheiro, R., Tømte, C., Barman, L., Degn, L., & Geschwind, L. (Eds.). (2023b). *Digital transformation in Nordic higher education*. Palgrave.

Pinheiro, R., Tømte, C. E., Laterza, V., & Asante, M. O. (2023c). Digital transformations in higher education before and following COVID-19: A Scandinavian tale. In L. Leišytė, J. R. Dee, & B. J. van der Meulen (Eds.), *Research handbook on the transformation of higher education* (pp. 366–381). Edward Elgar Publishing.

Reale, E., Avramov, D., Canhial, K., Donovan, C., Flecha, R., Holm, P., Larkin, C., Lepori, B., Mosoni-Fried, J., Oliver, E., Primeri, E., Puigvert, L., Scharnhorst, A., Schubert, A., Soler, M., Soòs, S., Sordé, T., Travis, C., & Van Horik, R. (2018). A review of literature on evaluating the scientific, social and political impact of social sciences and humanities research. *Research Evaluation, 4*(27), 298–308.

REF. (2021). Research excellence framework. Retrieved March 29, 2023, from https://results2021.ref.ac.uk/impact

Ritchie, H. (2020). Sector by sector: Where do global greenhouse gas emissions come from? Retrieved July 1, 2024, from https://ourworldindata.org/ghg-emissions-by-sector

Ritchie, H., Rosado, P., & Roser, M. (2020). Breakdown of carbon dioxide, methane and nitrous oxide emissions by sector. Retrieved June 7, 2024, from https://ourworldindata.org/emissions-by-sector

Römgens, I., Scoupe, R., & Beausaert, S. (2020). Unraveling the concept of employability, bringing together research on employability in higher education and the workplace. *Studies in Higher Education, 45*(12), 2588–2603.

Satell, G. (2017). The 30 year rule: How long does innovation take? A Blog in Disruptor League. https://www.disruptorleague.com/blog/2017/10/06/the-30-years-rule-how-long-does-innovationtake/

Schofer, E., Ramirez, F. O., & Meyer, J. W. (2000). The effects of science on national economic development, 1970 to 1990. *American Sociological Review, 65*(6), 866–887.

Selznick, P. (1957/1984). *Leadership in Administration a Sociological Interpretation.* Harper and Row.

Sin, C., Veiga, A., & Amaral, A. (2016). *European policy implementation and higher education. Analysing the Bologna Process.*

Smeby, J.-C., & Stensaker, B. (1999). National quality assessment systems in the Nordic countries: developing a balance between external and internal needs? *Higher Education Policy, 12*(1), 3–14.

Sørensen, M. P., Geschwind, L., Kekäle, J., & Pinheiro, R. (2019). *The responsible university: Exploring the Nordic context and beyond.* Palgrave Macmillan.

Spangenberg, J. H. (2017). Hot air or comprehensive progress? A critical assessment of the SDGs. *Sustainable Development, 25*(4), 311–321.

Spira, F., Tappeser, V., & Meyer, A. (2013). Perspectives on Sustainability Governance from Universities in the USA, UK, and Germany: How do Change Agents Employ Different Tools to Alter Organizational Cultures and Structures? In S. Caeiro, W. Filho, C. Jabbour, & U. Azeiteiro (Eds.), *Sustainability Assessment Tools in Higher Education Institutions* (pp. 175–187). Springer.

Stensaker, B., & Harvey, L. (2011). *Accountability in Higher Education: Global Perspectives on Trust and Power.* Routledge.

Stensaker, B., & Hermansen, H. (2023). *Global, Nordic, or institutional visions?* An investigation into how Nordic universities are adapting to the SDGs. https://link.springer.com/article/10.1007/s10734-023-01047-3

Taalas, P. (2021). *Ilmastonmuutos ilmatieteilijän silmin* [Climate change through the eyes of a meteorologist]. Tammi.

The World Counts. (2023). Global challenges. Retrieved March 27, 2023, from https://www.theworldcounts.com/challenges/global-warming/global-co2-emissions

Trondal, J., Pinheiro, R., Keast, R., & Noble, D. (2022). *Governing complexity in times of turbulence.* In J. Trondal, R. Keast, D. Noble, & R. Pinheiro (Eds.), *Governing complexity in times of turbulence* (pp. 2–12). Edward Elgar Publishing.

Vedung, E. (1998). Policy instruments: Typologies and theories. In M. L. Bemelmans-Videc, R. Rist, & E. Vedung (Eds.), *Carrots, sticks, and sermons: Policy instruments and their evaluation* (pp. 21–58). Transaction.

Veiga, A. (2023). Unthinking the European Higher Education Area: differentiated integration and Bologna's different configurations. In C. Dienel (Ed.), *Globalizing higher education and strengthening the European spirit* (pp. 93–110). Routledge.

Watson, D., Hollister, R., Stround, S., & Babcok, E. (Eds.). (2011). *The engaged university: International perspectives on civic engagement.* Routledge.
Witte, J. (2008). Aspired convergence, cherished diversity: Dealing with the contradictions of Bologna. *Tertiary Education and Management, 14*(2), 81–93.
Witte, J. (2021). Higher education, science, and the climate crisis. In S. Marginson, B. Cantwell, D. Platonova, & A. Smolentseva (Eds.)., *Assessing the Contributions of Higher Education.* Edward Elgar Publishing. Open Access. Retrieved March 27, 2023, from https://www.elgaronline.com/edcollchap-oa/book/9781035307173/book-part-9781035307173-14.xml
World Nuclear Association. (2022). Nuclear energy and sustainable development. Updated October. Retrieved March 27, 2023, from https://world-nuclear.org/information-library/energy-and-the-environment/nuclear-energy-and-sustainable-development.aspx
Young, M., & Pinheiro, R. (2022). The Post-Entrepreneurial University: The case for resilience in higher education. In R. Pinheiro, L. Frigotto, & M. Young (Eds.), *Towards resilient organizations and societies: A cross-sectoral and multi-disciplinary perspective* (pp. 179–200). Palgrave.
Young, M., Pinheiro, R., & Avramovic, A. (2024). Unpacking resilience in higher education: Investigating twenty-first-century shifts in universities' academic cores. *Higher Education, 87*(1), 221–236.

CHAPTER 2

Climate Change Policy for Higher Education? Investigating the Cases of the United Kingdom and the Netherlands

Jouni Kekäle, *Evandro Coggo Cristofoletti*, *and Rómulo Pinheiro*

Introduction

The expected roles of higher education institutions (HEIs) in society are constantly evolving. The overall direction for the change appears to be rising, with increasing new roles and expectations alongside traditional duties. Writing in the late 1990s, Burton Clark (1998), a prominent figure

J. Kekäle (✉)
School of Educational Sciences and Psychology, University of Eastern Finland, Joensuu, Finland
e-mail: jouni.kekale@uef.fi

E. C. Cristofoletti
School of Applied Sciences, University of Campinas, Campinas, Brazil
e-mail: evcoggo@unicamp.br

R. Pinheiro
Department of Political Science and Management, University of Agder, Kristiansand, Norway
e-mail: romulo.m.pinheiro@uia.no

J. Kekäle, R. Pinheiro (eds.), *Higher Education Policy for Tackling Climate Change*, https://doi.org/10.1007/978-3-031-78479-8_2

in the applied field of the sociology of higher education (HE), foresaw internal (e.g., cumulative and exponential growth of knowledge, new ideas and fields, and advances of technology in knowledge production) and external drivers (e.g., new and rising expectations from society, students' increased expectations for quality) for the observed changes: as a result, demands exceed resources and the HEIs' capacity to respond on every level. Even though Clark (1998) carried out this analysis two and half decades ago, the dynamism between society and universities, and in particular the range of demands and expectations set for the latter have intensified. The overall direction remains: more tasks and expectations are directed at HEIs, as documented by Enders and Boers (2009), with external pressures for demonstrating societal impact becoming increasingly salient (Sørensen et al., 2019).

In addition to increasing competition for knowledge production, faculty, and students, one of the most recent and fundamental additions to the external expectations for HEIs is the United Nations' (2015, 2022) 17 Sustainable Development Goals (SDGs). SDGs reflect universal challenges to humankind: it would be difficult to argue that they are not important. However, at the same time they hide the specificities of the ways in which problems manifest themselves and create conflicts with other local agendas. In addition, they do not directly address serious problems such as crime and violence or other types of socio-economic externalities. Yet, they signal global policy intentions towards tackling important wicked problems, like Climate Change (CC), without prescribing how to do so and who should be in charge.

According to the UN (2022), HEIs everywhere in the world must adopt a stronger and more active role to tackle the world's most pressing problems, along the lines of three interrelated key themes:

1. A move towards inter- and transdisciplinary modes of knowledge production and circulation.
2. The imperative of open science, becoming increasingly open institutions.
3. The demand for a stronger societal impact and much more proactive outreach activities from HEIs, "that allow for science advice for policy, advice on remedies or solutions, and engagement in societal projects oriented towards improving ecosystems, combating climate change and generating greater welfare in society" (UN, 2022, p. 84).

In Europe, HEIs are, for the most part, politically steered; the basic steering mechanisms being legislation and funding allocations either directly (basic grant and/or performance-related) or in the form of competitive tendering. Kaldeway (2018) has pointed out that, as global "grand challenges" SDGs have quickly become a topic central to HE policy and funding schemes everywhere. HEIs, in turn, have reacted to this change by adjusting (in various degrees) their strategies, internal structures, and primary tasks around the promise to tackle such grand challenges (Nowell et al., 2020). One aspect to be noted pertains to the fact that, in essence the SDGs are somewhat broad and need to be translated or contextualised to fit local circumstances, including the specific national systems and regional geographies underpinning the activities of HEIs.

To the extent that institutional promises to solve or tackle grand challenges such as CC—our sole focus within the scope of the UN SDGs, given the remit of this volume—have been made by many European HEIs and dedicated governmental funding streams (e.g., UN SDG, 2023), there are legitimate expectations that HEIs should do more about solving the complex and wicked CC problem. Rickards and Pietsch (2022, p. 1) contend that CC ranks as "the most important mission for universities of the 21st century". It appears to us that the development described above forms a basic tension in contemporary HE policies. Some potential consequences of this tension were highlighted in a recent survey where 68% of 400 global academic leaders, researchers, and heads of funding bodies surveyed agreed that the inability to communicate the benefits and impact of research for the wider society could become an existential risk to the whole academic enterprise (Dixon, 2023).

Given this atmosphere, it is perhaps no wonder that there appear to be different assumptions behind national steering policies concerning the role of HEIs in addressing societal challenges such as CC: are HEIs seen predominantly as knowledge providers for responsible policymakers and the industry (with their responsibility to diminish their own emissions and to demonstrate impact), or is their role seen as that of a societal coordinator or a policy implementer of measures for the mitigation of CC?

Given these important deliberations, the research question posed in this chapter is thus: *what are the official national policy approaches for steering HEIs towards CC, and how do two contrasting European HE steering systems (the Netherlands and the UK) navigate the tensions stemming from increasing stakeholder expectations together with the internal challenges of HEIs to address them?* In addressing this question, we begin with the observation or assumption that each national government has a legitimate role to steer HEIs and other organisations via legislation, funding mechanisms, and established agreements or procedures. This steering might remain vague (loose) or be more concrete (tight), but the potential for steering HEIs towards particular policy goals exists. Hence, we describe empirically the national steering philosophies, approaches, and their backgrounds through the two cases.

We approach the above-mentioned question by discussing two different central national policy approaches to deal with the tension of CC in relation to HE. Thus, the chapter primarily deals with official national steering policies in the way that national governments steer and incentivise HEIs in the mitigation of CC, including the official national role set for the HEIs to deal with the problem. In other words, we ask: how have the national government–HE pacts or agreements been settled in two different European counties: the Netherlands and the UK? Our intention is not, however, to compare or describe individual institutions and their actions in CC mitigation, as this is addressed in other chapters in this volume, like Chapter 5 by David Charles.

Social impact involves making a positive difference or improvement in society. From a historical perspective, the overall quest for a better societal impact, quality (viewed from the stakeholders' perspective), and relevance has continued for decades. Quality movements started in HE in the 1980s, and gradually moved towards the notions of the responsive and responsible university, the civic university, and Open Science, to name just a few trends (Kekäle & Pinheiro, 2025). More recently, a new international discussion geared towards research assessment has become rather prominent in the Global HE policy/governance landscape (DORA, 2012/2024; Pietilä et al., 2023). A rather broad international view is that research assessments concentrating on publication lists, H-indexes, and Journal Impact Factors have steered research towards intra-academic reputation games (DORA, 2012; Kekäle et al., 2023), and this incentive structure has hindered researchers from demonstrating and achieving quality and societal impact—for example, with respect to CC mitigation. Numerous international statements related to the theme, such as the San Francisco Declaration on Research Assessment (DORA, 2012) and the Leiden Manifesto for Research Metrics (2015),[1] basically aim at ending the supremacy of using bibliometrics as a way of guiding science and raising

[1] FOLEC-CLACSO in Latin America.

the role and significance of qualitative assessment together with responsible metrics (i.e., ethical and appropriate use of quantitative and citation-based metrics). Lately, the focus of EU science policy has been on the renewal of researcher assessments (CoARA, 2023). The movement is intended to be international. Researcher assessments are a key steering mechanism towards the future of HEIs and, arguably, towards better impact. Assessments are fundamental when deciding what kind of research is to be funded, who will be promoted, and what kind of academics are recruited: in short, assessments shape the future of HEIs. To provide some background for our policy discussion, we start by describing the recent transnational (EU-wide) CoARA (Coalition for Advancing Research Assessment) initiative, which can be seen as an attempt to foster the new, broader roles of HEIs throughout Europe and elsewhere in which HEIs would (be able to) provide better impact and relevance for society, including tackling CC (Kekäle et al., 2023).

The chapter is organised as follows. Next, we shed light on one of the key approaches in international HE policy by discussing persistent and recent movements towards the expectation set for HEIs to improve social impact, also in CC policy. We continue by discussing HEIs' potential roles in CC mitigation. Following that, we turn to a comparison of two national cases, the Netherlands, and the UK, which demonstrate different approaches to HEIs' expected roles in CC mitigation. Discussion and conclusions finish the chapter.

CoARA: Towards a Better Social Impact and Quality

Several recent studies have argued, somewhat controversially, that modern HEIs should be measured by their ability to tackle global challenges and to contribute to impact agenda, rather than on the basis of academic and scientific production per se (Waas et al., 2010; Findler et al., 2019; Klofsten et al., 2019; Filho et al., 2019; Corazza & Saluto, 2021). Cottafva et al. (2022, p. 138) suggest: "While bibliometric tools for the SDGs are currently emerging, the existing models have not focused on interdisciplinarity or on their usefulness as decision-management tools to drive SDG-related research".

Established in early 2022, CoARA aims at, among other issues (e.g., methodological aspect and intra-scientific quality), recognising the diversity of contributions to and careers in research, so that research assessment would better reflect the new and broader societal roles in academia. One of the core ideas appears to be that through different recruitment and assessment criteria a broader set of competencies would flourish in academia: competencies that deal with open science, impact, and societal relevance (CoARA, 2023; Sivertsen & Rushford, 2022).

The joint initiative to change the research evaluation system, exemplified in CoARA (2023), originates from several research funders and is backed by the European Commission. Key European science policy actors[2] have also committed themselves to the project extensively. As of November 2024, the agreement has been signed by 805 HEIs and international organisations, including institutions from the UK, Canada, Australia, Brazil, and Chile.[3] Behind this is a high-level political decision and extensive preparation by the Commission. The change in the research evaluation system and research careers are part of the European Research Area (ERA) agenda and actions (cf. De Elera, 2006), which the EU member states have committed themselves to. As the basic task of research funders is to allocate money for research (agenda setting), questions of quality, effectiveness, and social relevance play a central role. Participation in the CoARA consortium is voluntary and participants can leave the consortium at any time. This is a major international attempt to enhance HEIs societal impact, but it is not the only. The UK's Research Excellence Framework (REF) (to be discussed in the next section) is another.

As noted, the overarching goal of CoARA has been to maximise the quality and impact of research (CoARA, 2023). While this remains the ultimate goal, academic freedom, multiple societal needs, and the independent status of HEIs tend to diversify the concrete strategic aims and targets. Consequently, CoARA's somewhat indirect focus on renewed assessments (ultimately aiming at better quality and societal impact) has resulted in several working groups discussing the diverse aspects of assessment. One of the nine working groups listed on the CoARA pages (CoARA, 2023/ working groups) directly aims at Transdisciplinary, Applied/Practice-Based Research, and Impacts, and mentions CC as a case in point.

CoARA was initiated in the Netherlands and the country has been a forerunner in the CoARA movement ever since. The Netherlands has a strong tradition of research on CC since the 1960s. The work has

[2] These include, among others, the central European funding organisation Science Europe, European University Association, the EU, and many research funders.

[3] CoARA. Signatories. Available at: https://coara.eu/agreement/signatories/

produced hundreds of projects and publications on the topic (Veraart et al., 2014) with many key HEIs taking an active strategic role. Since many HEIs in the country—and indeed around the world—are engaged in efforts to tackle CC (Filho et al., 2023), the improved strategic impact should also be targeted at the mitigation of CC. After all, the situation around the global CC is getting worse, and the problem is urgent. For the Netherlands, an additional problem is that a major part of the country's land area is below the sea level.

This brings us back to the SDGs and CC, and other societal impact challenges, which are among the implicit key targets of the CoARA and other similar joint agreements aiming at increasing societal impact. Although it is not explicitly mentioned, there appears to be an implicit view that universities and research have been siloed, and that the promised strategic solutions to issues such as CC and the societal relevance of research, are obviously still lacking (DORA, 2012; Kekäle et al., 2023). If researchers and funders operate on the basis of expert information, numerical data, and citation indexes, laymen and non-experts—the public—are somewhat left behind. According to this line of thinking, research does not radiate to society at large and other societal branches in the way desired, or promised, as scholars play their 'internal game' of publish or perish. Society is financing research activities, and global challenges such as CC require cooperation on many fronts (without discounting the role of basic academic research). As a result, policymakers expect HEIs to react accordingly by accommodating these new demands into their existing (and new) structures, tasks, and networks.

HEIs' Potential Role in Mitigating Climate Change

CC is arguably the most crucial problem of our times, and from a global and western perspective it is perhaps the key grand challenge that humanity has ever faced. Cook (2019) has remarked that 97% of scientists agree that CC is real and caused by humans. Countering misinformation requires a multi-disciplinary approach, cultural values, individual cognition, societal trends, and developing technology. A changing media landscape (from traditional, professionally curated to personalistic and fast-paced social media) contributes to the multi-faceted problem.

Global warming, or CC, is caused by increasing greenhouse gas emissions that cannot be absorbed by soil and natural sinks. CO_2 emissions are rising mostly because of the fossil fuels burned during energy production.

The energy sector (electricity, heating, and transport) is responsible for no less than 73% of carbon oxide emissions in the EU alone (Chap. 1). In the 2016 Paris Agreement, the EU committed to reduce its greenhouse gas emissions by at least 40% by 2030 in comparison to the 1990 level. In 2021, the goal was tightened to a reduction of at least 55% by 2030, reaching climate neutrality by 2050. Effective actions need to take place quickly, within a few decades (Levin et al., 2012).

Witte (2021) has pointed out that science and research can make three major contributions towards tackling the climate crisis, namely: (1) by measuring and understanding CC to provide an empirical basis for policy decisions and political action; (2) by developing climate-friendly technologies and economic and social models that allow CC to remain at bearable levels (mitigation); and (3) by helping to devise coping and adjustment strategies enabling humankind to deal with CC (adaptation).

Witte's (2021) first type of contribution is based on the traditional roles in which HEIs have been rather successful since CC is acknowledged by most people, despite existing skeptics tied to political and economic agendas (right wing populism, industry lobbyists, etc.). Recent (2023) data from the UK's Research Excellence Framework (REF[4]) show examples of other types of contributions. However, innovations by HEIs and other actors typically require decades before they can be utilised in a large scale and have measurable impacts. This applies especially to major systemic changes such as changing national energy production and transport.[5] Satell (2017) refers to the 30-year rule that means that humankind largely has to rely on existing solutions although political incentives and industrial solutions are crucial.

There is a general consensus that there already exist many technical solutions for the green transitions (society and economy): wind, solar, nuclear, and water energy, heat pumps, CO_2 capture and storage, plus a hydrogen economy. HEIs can directly contribute via education, supply of qualified workforce, attitudes towards sustainable lifestyles, improving processes, and reducing emissions from their own operations (Findler et al., 2019). Moreover, HEIs can innovate and promote start-ups and business incubators. However, HEIs cannot directly invest in existing green technologies in industry on a major scale, nor can they set any rules

[4] Online at: https://results2021.ref.ac.uk/impact

[5] COVID-19 vaccine development was much quicker, but the implementation was delivered via vaccinations.

and regulations in societies that would directly mitigate the problem in the crucial energy sector. The key role here would need to be taken by politicians and the industrial sector at large. It would therefore appear to be a mistake if HEIs are seen as the primary actors responsible for solving the problem of CC (e.g., Grau et al., 2017, p. 38). In other words, the primary role for HEIs in this respect is to perform a *developmental* or supportive role—as identified by Witte (2021)—rather than a generative one as the primary engine for CC mitigation (Gunasekara, 2006).

Methods, and Background for the Two Cases

The adopted methodological case-based approach (Yin, 2003) is a comparative one, investigating developments across two relatively distinct European countries and HE systems: the Netherlands and the UK. In addition to representing two rather distinct HE systems in terms of size, institutional profiles, and scope of activities, these two cases are fundamentally different in terms of the society–HE pact as to the HEIs' expected role in the mitigation of CC. The approaches and policy steering ideologies appear to be linked to the broader issue of the HEIs' impact and relevance. To learn more about the global/national climate change agenda and the impetus for a stronger contribution from the HEIs, we started by analysing an existing database. Our empirical account started with an interest in the UK's REF, which provides an open database for exemplifying the impact of HE research. Each REF submission includes impact case studies describing the specific societal impacts that occurred during the assessment period (1 August 2013 to 31 July 2020) and were underpinned by excellent research undertaken in the submitted unit. The REF is run by Research England (RE) which is funded by the central government but staffed by people from the university sector to ensure that decisions are not politically driven. Government and RE discuss the objectives of policy, and RE ensures it is implemented in a fair manner.

Following de Boer et al. (2008), the governance of the two higher education systems under study has changed over the course of the twentieth century. The UK went through three phases: the first one was marked by the dominance of academic self-governance—the government remained more distant, especially in the control and steering mechanisms. The second and third phases are still present to this day and represent a shift. The second phase included the strengthening of government intervention in the new universities and polytechnics, which then spread within the sector.

The third and most significant phase was the adoption of competitive pressures and market-orientated governance, influenced by the New Public Management reforms of the 1970s. In the Netherlands, there was a notable increase in the national government's involvement in universities after the Second World War. This phase was characterised by the implementation of extensive legislation, regulation, and administrative oversight, resulting in an arrangement where academic self-governance and governmental interference were closely connected. A significant shift occurred in the mid-1980s when the government embraced the concept of 'steering from a distance'. This shift marked a departure from strong regulatory approaches, emphasising a philosophy in which the government's role shifted towards establishing a broad framework for the university system, moving away from detailed regulation (Goedegebuure et al., 1993). Table 2.1 shows data on the institutional characteristics of HE in the case countries. It reveals that there are marked differences in institutional profiles amongst the national cases, with most research universities in the UK being private organisations with government funding, while in the Netherlands there is a strong public institutional base.

In addition, Table 2.2 below shows key R&D (research and development) indicators of relevance to the cases in question.

Our initial starting point was that by using the UK as a case study, we should be able to, first, investigate the extent to which the UK government has devised and/or implemented a series of (HE) policy initiatives in the realm of CC and, second, to highlight the policy–research interaction. A search in the REF 2021 database with the keyword *Climate Change* produced 637 hits, where it was mentioned as an example of research impact. REF can be seen as a policy tool, but the cases themselves do not necessarily show UK policy initiatives. Nevertheless, this initial analysis provided a good picture of the HEI sector's contribution to societal impact and the mitigation of CC.

In this initial analysis, we first analysed the 637 impact narratives or descriptions to gain a comprehensive picture of the topic and UK universities' role in it. To gain a picture of HEIs' impact on CC, approximately 100 of the most relevant studies were singled out on the following grounds: (1) aim at mitigating CC; (2) applicability to many countries, especially with the biggest emissions; (3) relatively direct policy contributions, models, tools, and attempts at solving or mitigating the root causes of climate change, especially in the energy sector generating the biggest part of emissions (in the UK their share of greenhouse gas emissions was 85% in 2018).

Table 2.1 Higher education in the UK and the Netherlands(2022)

Institutions	*N*	*Private*	*Private-government dependent*	*PhD awarding*
UK				
Alternative provider	95	95	0	0
University	137	1	126	127
University college	35	1	34	8
Total	257	97	160	135
Government expenditure per student, tertiary (% of GDP per capita)—World Bank data, 2016			38%	
School enrolment, tertiary (% gross), Word Bank data, 2021			77%	
The Netherlands				
Open university	1	1	0	1
University	19	11	7	18
University of Applied Sciences	37	1	35	0
Total	57	13	42	19
Government expenditure per student, tertiary (% of GDP per capita), World Bank data, 2016			35.8%	
School enrolment, tertiary (% gross), World Bank data, 2021			89%	

Source: Compiled based on ETER Report UK (2022), ETER Report Netherlands (2022), World Bank data

Table 2.2 R&D indicators for the UK and the Netherlands

R&D data	*UK*	*The Netherlands*	*Source*
Researchers in R&D (per million people), 2021 and 2019	4491	6074	World Bank
Higher education expenditure on R&D (HERD) at current prices and PPP$	13,177.9 (2019)	6643.3 (2020)	OECD Library
Research and development expenditure (% of GDP), 2021	2.91%	2.31%	World Bank
Percentage of GERD financed by higher education and PNP sectors, 2021	11.42%	2.47%	OECD Stat
Percentage of GERD performed by higher education and PNP sectors, 2021	22.52%	28.47%	OECD Stat

GERD: Following the recommendations of the Frascati Manual (2015), GERD is gross expenditure on research and development. It captures the total expenditure (current and capital) on R&D by certain sectors of the economy: resident companies, higher education, government, and non-profit institutions

It is important to note that the primary researcher (Jouni Kekäle), here responsible for reading and assessing the cases, can be seen as an informed citizen, but he is not an expert in all the disciplinary areas involved. His field of research is HE studies, with the disciplinary backgrounds of psychology and sociology. The sample and the overall impact cases appeared to display incremental improvements, significant scientific work, but no system-level solutions, game changers, major breakthroughs, or innovations, such as hydrogen economy. The university project would only be a small part, or it serves as the knowledge base needed for such major developments, but no doubt possible cooperation breakthroughs would have been reported in the case studies. We also carried out several further searches of the REF cases including 'Governmental policies climate change' which produced 17 hits. In most of the cases (14), the policy impacts of UK research were directed, for example, at developing countries rather than the UK as such.

Simultaneously with the UK CC policy analysis, another project (Kekäle et al., 2023) dealing with new assessment systems for researchers has been ongoing. This work on research assessment has involved long analyses and discussions on the backgrounds of the new approaches. It has revealed a philosophy according to which the new roles of HEIs to provide societal impact (e.g., mitigation of CC) could be organised. The background discussion related to these two studies raised our awareness to the rather fundamental differences in different countries' CC mitigation policies and the assumed role of HEIs at the national level. These analyses turned our attention to the interesting differences in HE policy related to CC in the UK (a non-EU country) on the one hand, and the Netherlands, a core EU member state in central Europe, on the other. As noted, the Netherlands is a forerunner in the CoARA movement and in promoting the new, more active role of the HEIs in providing societal impact. The country is also an example of a 'traditional' EU member state well integrated in matters related to EU Science Policy and can be seen as representative of the Central European EU countries more generally. The UK and the Netherlands, it turned out, hold different assumptions towards the basic tensions described in the beginning of the chapter, namely the HEIs' role in CC mitigation in the national policy. We discovered that a comparison between two diverse national cases might be useful in describing the changing nature of the 'social pact' between HE and society in relation to CC: the contrasting society–university pact brokered by the state (when it comes to the HEIs' expected role in mitigation of CC). In cultural studies, comparisons tend to bring out unique features that may help us to understand the assumptions and values that affect different ways of life (Kekäle, 2001). In short, we found out that there are different drivers, origins, and backgrounds of HE policy and steering, and different modes of the 'social pact' between HE and society in relation to CC. Next, we start from the Netherlands and then turn to the UK.

Empirical Findings

The Netherlands

Interestingly, the Dutch Ministry of Economic Affairs and Climate Policy is responsible for CC in political terms, but a national climate agreement distributes the actions and responsibilities to different actors, including HEIs. According to the central government, it "wants to reduce the Dutch greenhouse gas emissions by 49% by 2030, compared to 1990 levels, and a 95% reduction by 2050. These goals are laid down in the Climate Act on May 28, 2019. The Climate Plan, the National Energy and Climate Plan (NECP) and the National Climate Agreement contain the policy and measures to achieve these climate goals" (Government of Holland. Climate Policy, 2019). The Mission of the Ministry is to promote sustainable economic growth in the Netherlands, and the country has been able to reduce its greenhouse gas emissions by 18% between 1990 and 2019 (National Inventory Report, 2021).

The Netherlands has chosen a distinctive consensual and collaborative national strategy in the face of CC: a Climate Agreement, containing policy and measures, between 100 key societal parties, was completed in the Hague on 28 June 2019. The starting point for the agreement was that as the principle goal of reducing greenhouse gas emissions, with the outcomes affecting the everyday life of all citizens across all sectors and businesses. The national challenge and transition were primarily seen as a *social* transition. The Netherlands was striving to achieve more ambitious emission targets by 2030 than the EU. The Dutch Climate Agreement aimed at containing a package of measures which has the broadest possible societal support, and as many contributing parties as possible to reach the reduction targets.

This approach required an active societal role from the HEIs in the mitigation of national CO_2 emissions. Domestic HEIs are set to actively contribute to reductions in the Dutch real estate sector, and HE and university education are seen as a part of the solution not only in the usual technical sense of reducing emission from the university buildings, but also generating climate-related competences and knowledge in professional education. This link is stronger in the Agreement's section 'Labour Market and Training of the Agreement', where education, training, and influencing peoples' mindsets were tightly incorporated in the agreement.

> Climate-smart production should be an integral part of the programmes in the field of livestock farming and plant cultivation at all levels (senior secondary vocational, higher professional and university education) and will be embedded in greater depth in the development of the existing GroenPact platform. (Climate Agreement, 2019, p. 131)

Training and knowledge transfer for innovation were emphasised, with "sufficient influx and turnover of talent and the translation of scientific knowledge into practical applications in the business community, for example through climate-proof programmes, practice-oriented research and learning communities" (Climate Agreement, 2019, p. 131), alongside the embedding green skills in lifelong learning programmes through education. It was also seen as important to create awareness of the urgency of CC in young people, and various skills and competences were agreed to be developed and implemented through cross-sector cooperation—much in the spirit of the UN's (2015, 2022) SDGs.

Education and training were given great importance in the agreement, which was substantiated in the strategic motto 'People make the transition'. Since 2016, the Ministry of Education, Culture and Science has been carrying out experiments in higher professional education aimed at advancing tailored solutions in education for adults. Education and educability were trusted to generate change (Climate Agreement, 2019, p. 218). Digitalisation, Information and Communications Technology (ICT), and educational innovations were fostered in the realm of university education.

Governance Policy and HEIs' Response

In line with the developments, the Netherlands is the country leading the CoARA movement, with a historical development work along the similar lines, as will be shown, and the chairperson of the CoARA also originates from the Netherlands. Already in 2019, the country launched a prior initiative titled 'Room for everyone's talent' by the universities of the

Netherlands and leading Dutch funding organisations. Also, the term 'Recognition and Reward' has been used to signify the movement. The initiative (ongoing) and the subsequent movement aim at reaching a new balance in recognising and rewarding the work of the academics: in academic assessments less emphasis would be placed "on the number of publications, and a greater emphasis on the other domains in which the academic is active, such as education and impact" (Room for everyone's talents, 2019). The Dutch Research Council (NWO), one of the key initiators, is one of the most important science funding bodies in the country aiming at fostering quality and innovation in science. The wider variety of academic careers, paths, and profiles would result from the national initiative. Through different assessments and recruitments, the core aim is to foster both leadership roles and societal impact.

The movement is very much in line with the UN's (2022) expectations of the contributions and solutions by HEIs across the world to the SDGs. As a result, key Dutch HEIs including the universities of Twente, Maastricht, Leiden, Amsterdam, Tilburg, TU Delft, and Utrecht are already working towards changing academic profiles—so that some academics might concentrate on providing societal impact, for example. Each of them has signed the CoARA agreement. Currently (in 2023/2024) Maastricht holds the chair of CoARA. Nearly all 21 (30 in 2025) Dutch research universites, funding bodies and supoor organisations are signatories. Among signatories are many key stakeholders, including domestic funding agencies (CoARA, 2023).

Dutch universities have progressed the aims of the room for everyone's talent scheme and the subsequent CoARA movement in their own ways but aim at similar outcomes. For example, the strategy of TU Delft (2021) aims at creating 'Impact for a better society' under the terms 'Recognition and Reward' by diversifying and vitalising career paths. Valorisation (the process of creating societal and economic value from knowledge) is supported by finding people's strengths, diversifying careers, use of narrative assessments, rewarding teamwork, and so on. For the researchers involved, therein lies a promise that their diverse talents are to be recognised and rewarded; for society, a better form of impact will result from the strategy (TU Delft, 2021).

There is a distinct national policy behind the strong faith in HEIs role in the mitigation of CC. However, not all domestic scholars have been convinced of the developments in the Netherlands, as there is tension between 'pure science' and the socially driven agenda. In their open letter,

171 established academics, especially from the hard sciences (STEM), warned that the new Dutch assessment system (Room for everyone's talents) and its emphasis on social relevance will lead to more arbitrariness and less quality, and that this policy will have major consequences for the international recognition and appreciation of Dutch scientists. The critical academics stated that Open Science criteria are not scientific, but rather political. According to them, the model will have particularly negative consequences for young researchers who will not be able to compete internationally. In response to this critical open letter from the older scientists, 113 younger scientists defended the proposed changes. In an open letter, they stated that scientists nowadays do more than research: "Therefore, in our view, scientific publication is no longer the only unit to express quality, because it is not representative of the tasks of the modern scientist" (Recognition and Rewards: panel discussion, 2021).

The UK

In the UK, as elsewhere, HEIs are under increasing pressure to demonstrate their relevance and contribution to society, offer improved 'value for money', and provide students with skills and experience when entering the job market (Mbembe, 2016). However, we can observe many differences when the UK is compared to the Dutch approach towards CC in HE Policy.

The UK has adopted emission targets based on the long-term temperature goal of the Paris Agreement (Progress Report to Parliament, 2022). The Climate Change Act of 2008 in the UK introduced a legally binding target by the year 2050 to reduce greenhouse gas emissions by at least 80% when compared to 1990 levels. In 2019 the target was tightened on achieving net zero emissions (100%) by 2050 (Cran-McGreehin, 2022). This is a world-leading target. So far, the country has made strong progress in reducing emissions, between 1990 and 2017 by 42%, while the economy has grown by more than two thirds. In terms of historical cumulative emissions, the UK has indeed been able to greatly diminish the output of carbon dioxide into the atmosphere.[6]

[6] Global industrialisation started in the UK. Therefore, the country produced around 60% of global emissions in the mid-1700s–1800s. Today, the UK's total emission equals to 4.25% of the global cumulative emissions (Our World in Data 2023). A major reason for sinking emissions in the UK is the structural change in the economy. The UK's economy is currently dominated by services industries. These include retail, hospitality, professional services, business administration, and finance, which do not produce huge amounts of emissions. Heavy

The UK has taken a legal or legislative road with the more traditional role of relatively independent universities, rather than the national agreement approach as in the Netherlands. There is no explicit demand for HEIs to engage in research on CC; the demand is more diffuse and comes mainly from the research community itself and the funding agencies. HEIs are expected to provide CC education and to reduce their carbon emissions. For example, the University of Edinburgh has committed to the goal of becoming zero carbon by 2040.

As of December 2024, 17[7] out of some 140 UK universities as well as some professional associations had signed CoARA. That equals 6.6% of the nation's universities. However, of the signatories only the Universities of Edinburgh and Warwick are among the ten most highly ranked UK HEIs. Other major HEIs (e.g., members of the prestigious Russell Group) were not signatories as of the time of writing, December 2024 (CoARA, 2024). In a way, the UK situation is quite opposite to that in the Netherlands.[8]

Policy Framework: Climate Change Committee and the REF

With the passing of the 2008 Climate Act, a strong independent body called Climate Change Committee (CCC) was established to assess the overall developments in CC across the UK. The CCC is composed of a chair and eight members appointed by the UK government. The members provide a wide range of expertise with backgrounds from across academia, civil society, and business to develop and publish advice and recommendations concerning climate change policy.

It can be interpreted that the CCC has acted as a buffer institution or mediator between the government and the HEIs (Coggo et al., forthcoming). According to domestic observers, the CCC has been "critically important to ensure consistent policy delivery and evidence-based

manufacturing has largely moved to developing countries such as China with cheaper labour costs. The structural change applies to many other developed countries as well.

[7] These include, amongst others, Nottingham Trent University, University of Reading, University of Derby, Queen Mary University of London, Loughborough University, University of Central Lancashire, University of Edinburgh, University of Strathclyde, Swansea University, and University of Warwick.

[8] The difference compared to the Netherlands is in line with the overall national policy. However, it can well be that we will witness more signatories from the UK as the movement gains political strength and more funding is connected to the key targets. After all, signing the initiative is relatively easy (symbolic compliance). Here our interpretation deals with differences in national policies.

decision-making" (Fankhauser et al., 2018, p. 1). Both the CCC and REF appear to display—and lead to—a distinct interplay between the government and the HEIs in the UK context, if compared to the Netherlands. Although no such interplay is free from policy imperatives, it nonetheless appears to reflect different cultural values and normative assumptions.

Much of the UK's impact analysis on CC is delegated to CCC. The independent annual report (Climate Change Committee, 2022) is written by university-based scientists. The CCC produces a Climate Risk Independent Assessment, setting out the risks facing climate change in the UK and providing advice to the government (Progress Report, 2022). For example, the CCC had previously indicated that the UK should aim at net zero emissions by 2045–2050 (Cran-McGreehin, 2022). The purpose of the CCC is to advise the government on emission targets and report to the Parliament on progress made in reducing emissions. The structure differs from those in many other countries. The law set out a politically independent body to assess the progress, which reports to the government that provides clear responsibilities for different political organisations and ministries (Climate Change Committee, 2022). The CCC gathers research evidence internationally, not only from the UK (Cristofoletti et al., forthcoming). In addition, the CCC has devised an instrument termed the 'call for evidence' where all organisations and researchers can respond and upload scientific evidence in their respective areas of expertise.

In 2023, our analysis of CCC and its work (including reports to Parliament) revealed that the HEIs are either often mentioned nor are they given a direct and central role in actions aiming at mitigating CC.[9] The key national informative and steering document, the annual CCC Progress Report (2022), displays only local and piloting approaches from the HEIs rather than a general attempt to change and broaden the roles of the HEIs as is situation in the Dutch case. CCC specifies mainly cooperation with the Universities of Lancaster, Sussex, and Aberdeen (CCC Progress Report, 2022).

Another influential Policy Paper by the UK government on CC, titled 'Climate change adaptation: policy information' (2022), refers to two domestic HEIs only. The official national CC policy approach towards HEIs is that these should inform policymakers (the parliament) about

[9]This reflects the official national steering policy in CC mitigation and what the CCC focuses on. Universities still have to reduce *their own* emissions dramatically based on legislation.

different policy choices regarding CC; in the official policy and steering, HEIs are predominantly seen as knowledge providers, not as buffers or policy operators implementing the policies set nationally by the government. The UK universities policies are guided by the Paris Climate Agreement (UNFCCC, 2015) where the UK played a key role.

The actions and accountability, however, mainly rest on the political and business sectors where the actions are also taken. This is seen in such statements in the report as "Business will need to undertake and report on an annual due diligence exercise on their supply chains" and the "UK Net Zero Strategy is highly reliant on productivity measures to reduce emissions from the agriculture sector" (CCC Progress Report, 2022, pp. 305–309). One commentator noted that the government's CC strategy "had a strong emphasis on what schools can do and was comparatively light on universities" (Tam, 2022, p. 1.)

As noted above, HEIs cannot play a key role in investments or enact fossil-free energy production legislation (a sector which in the case of the UK provides 85% of the greenhouse gas emissions, and where working solutions are already present). The policy mechanism used, CCC's reports to the Parliament, seem to direct accountability towards those political sectors and actors who play a direct role in green energy or the production of emissions. HEIs' societal accountability in matters related to CC is handled via the REF, which displays the impact of research taken in UK HEIs. As noted above, the 2021 public database analysis provided 637 impact narratives on CC. UK HEIs are also responsible for their own societal accountability through initiatives such as the civic university movement, so in that sense the REF is only partially representative of what the universities are doing (see Chap. 5 in this volume for a detailed account).

A More Varied Picture Underneath (or Alongside) the Official HE and CC Policy

In terms of national steering or, rather, influencing research, citizens, and HEIs in issues related to CC, there are other various policy documents. These include the Climate Change Act 2008, the Net zero 2050, the autonomous initiatives by the universities (including ranking, Russell Group, Universities UK) and funding agencies that compose a scenario of creating demand for research on CC.

UK HEIs have committed themselves to the fight against CC in a pamphlet (Confronting..., 2021) where different HEIs display their actions. The main argument is that UK HEIs are at the heart of innovation and that, most importantly, they are excellent spaces for multi-disciplinary research and cross-sectoral thinking. The document emphasises joint societal discussion and cooperation in tackling CC, especially cooperation between the government and the HEIs. The proposed approach comes close to the Dutch national approach. Other initiatives include UK Climate Resilience Programme providing strategic funding for CC-related research, UK Climate Resilience Programme, and the UK Research and Innovation (UKRI), with stimulus to research on CC.

An analysis of the use of UK research funds has been provided by UKCDR (2021). The report concludes, among other things, that UK-funded research on CC and international development on CC mitigations have had an important role in understanding the opportunities, challenges, and trade-offs associated with the COP26 priorities. Further alignment of the funding on CC research would be needed (UKCDR, 2021, p. 8). Indeed, the Net Zero Policy in the UK, despite not demanding directly that universities undertake research on CC, considers research as something fundamental.[10]

According to a recent policy paper by the UK government, by 2023 there will be new requirements for further education teachers to build sustainability into the curriculum. The aim is that there should be more opportunities for people to train, retrain, and upskill to enter green careers. Professionals working in climate research and policy, from both science and non-science disciplines, should dedicate a proportion of their working time to provide help to teacher-led climate education initiatives (Sustainability, 2022). When it comes to bottom-up initiatives, a document entitled 'National Climate Education Action Plan' was published in 2021 by the University of Reading (Reading, 2021).[11] Amongst other aspects, the plan stresses that no "single organisation, or indeed government department, is able to take this agenda forward alone" and that only "by working together will real change be possible" (ibid.). Finally, the UK's Climate Commission for Higher and Further Education has devised

[10] More information on these is available on the following websites: https://www.ukclimateresilience.org/; https://www.ukclimateresilience.org/; https://www.ukri.org/news-and-events/responding-to-climate-change/; https://www.ukcdr.org.uk/wp-content/uploads/2021/04/01880-UKCDR-Climate-ChangeReport.pdf; https://www.theccc.org.uk/wp-content/uploads/2019/05/Net-Zero-The-UKscontribution-to-stopping-globalwarming.pdf

[11] Climate Education Action Plan (https://www.reading.ac.uk/planet/climate-education/climate-education-plan)

a climate action toolkit to prepare staff and students to cope, thrive, and move responsibly in a changing world. This is exercised through such means as equipping students with employable skills to protect biodiversity, working towards climate and social justice, and lowering and eliminating greenhouse gas emissions (Climate Action Toolkit, 2021)

Overall, by the year 2023 the UK's official CC mitigation policy structure and mechanisms appear to reflect a faith on autonomous HEIs and their established roles and activities in the fight against CC. Key accountability is placed on those who are responsible for emissions. CCC and REF are policy mechanisms that reflect this decentralised policy culture, and the system is characterised by a clear demarcation of roles and responsibilities. A governance structure of this kind, however, does not necessarily mean that there would be lower ambitions towards CC mitigation than elsewhere, but it signals the broader UK impetus for greater decentralization and wide societal mobilization beyond the state and its agencies.

Discussion

The empirical material sheds light on two contrasting policy approaches as regards the governance of CC within European HE.[12] The Dutch approach towards national CC policy regarding HEIs and their roles is largely based on joint efforts and, in many places, also the direct input of the universities, where HEIs are increasingly seen as policy actors and implementers. As noted already in the position paper "Room for everyone's talents" (2019, p. 1), there is a need for greater emphasis on domains other than research where academics are active including education and impact: "This broader form of recognition and appreciation is better suited to the current core tasks" of HEIs "and what society requires of these establishments" (Room for everyone's talents, 2019, p. 1). This emergent strategy has been concretised by attempts to change assessment practices where HEIs and researchers are measured by their ability to tackle global challenges, rather than by their academic and scientific production. In the introduction to this chapter we assumed that the international change in political steering, for example by the UN (2022) and research funders (Kaldeway, 2018), alongside HEIs' commitments to SDGs in their strategies, had initiated a push and pull towards new types of service roles

[12] By mid-2024 the Dutch national policy appeared to be changing as new far right government proposed that funding for research and science will be cut by €1.1 billion.

(aiming at better impact and relevance). The Netherlands is a country which, on the basis of the materials and practices analysed, reacted rather seriously to the call, and carried out strong measures along these lines, as shown above.

In this respect it is not surprising that CoARA is led by the Dutch HEIs themselves, as they have taken an active role in shaping the agenda-setting process both transnationally as well as nationally, as a sign of early adoption. There is, however, still more work to do, although nearly all Dutch HEIs have joined the CoARA. The aims of the movement seem to be changing. The basic values may not be shared by all operators, and there has been internal criticism. Signing the agreement does not directly solve SDGs or CC: A recent survey (n = 400) revealed that in the Netherlands only 64% of respondents agreed with the statement that 'research should at least aim to have an impact on the wider world', whereas in the UK 93% of the respondents agreed with the statement (Back to Earth, 2023). In the long run, and given earlier developments in other domains, we might expect this leading group of early adopters, combined with policy pressures to respond, embrace or join CoARA, in what is often termed as 'isomorphic effects' (Ramirez et al., 2016), in this case of the mimetic and regulative types.

In contrast, the UK approach towards HEIs' role in CC is based on independent scientific analysis and follow-up, and on clear annual responsibilities set for political actors and business sectors causing emissions. HEIs have overall accountability on their own emissions and impact, as all organisations have, based on CC legislation, and many HEIs conduct scientific work on and around CC: yet the dominant legislative approach differs substantially from the one adopted in the Netherlands. Up to this point in time, UK HEIs have not been *pushed* to embrace similar *nationally prioritised* societal service or policy implementer roles despite ongoing policy and academic discussions concerning the civic/engaged roles of UK HEIs that have been ongoing for some time (Benneworth & Charles, 2012; Goddard et al., 2016). There is encouragement and *pull*, however, as the UK government awards block funding specifically for business and community engagement. There have also been initiatives to encourage greater engagement with the local communities. It seems that the preferred solution in the national CC policy has been to delegate the responsibility to an intermediary body, the CCC, which in essence acts as a *boundary organisation* (Hoppe et al., 2013) between (global and national) HEIs on the one hand, and science and national policy frameworks (agenda setting) on the other. These findings are aligned with earlier studies, suggesting that the CCC "functions primarily as a *knowledge broker*, offering trusted information to policymakers, and at times as a *policy entrepreneur*" (Averchenkova et al., 2021, p. 1218; emphasis added). Other studies suggest that, over time, the CCC has become more daring in challenging the policy status quo (Dudley et al., 2022).

That said, the data also revealed some similarities. Both countries share the nearly global key mission of reducing greenhouse gas emissions as the primary mechanism for addressing CC. In the UK, both the CCC (representing HEIs) and the Parliament (representing government and society) jointly monitor developments towards the set CC targets, whereas in the Netherlands the Ministry of Economic Affairs and Climate Policy functions as the central agency responsible for CC in political terms, with overall operations, policy, and measures set out in a national Climate Agreement (2019) and distributed to a variety of stakeholders, including HEIs.

We have noticed that the picture is more complicated and in all democratic systems different views exist side by side. The picture may also change; our chapter portrays the situation as of late 2023. However, if we simplify the picture, it could be argued that as a *transnational governance system* (Djelic & Sahlin-Andersson, 2006) CoARA is centred on leadership, delegation, and accountability with the European commission and national governments as 'leading' parties in steering (European) HEIs to address CC (and SDGs and societal impact more generally), as part and parcel of the broader social pact between HE and society, brokered via the state (Schwartzman et al., 2015). In the UK model—with their long industrial and university history—it seems that HEIs have retained more of their traditional roles in research and instruction (see Chap. 5 in this volume, by Charles). Education is a central element in both the UK and the Netherlands as regards the CC policy in and for HE.

Recent studies from the Nordic countries suggest that when faced with challenges associated with sustainability, the strategic response of Nordic HEIs has been either to aim for improving their market image or attempt to foster their competitive edge (impact) in terms of research and education (Stensaker & Hermansen, 2025). Similarly, a recent analysis of strategic responses by UK and Brazilian HEIs (Cristofoletti & Pinheiro, 2023) to the sustainability challenge (SDGs agenda) suggests that in both cases, the SDGs are used as strategic tools to try to connect scientific relevance with social impact; in other words, trying to resolve tensions between the two different pressures—how to produce research of excellence and impact at the same time.

Conclusion

This chapter compared two contrasting governance approaches with respect to HE policies and stakeholders' expectations by paying particular attention to the roles of HEIs in CC mitigation. In the Dutch system, in line with the national developments leading to the CoARA movement, HEIs are explicitly seen as having a central role in assisting the domestic and European economies and societies in moving towards a low-carbon scenario, in educating people under the title 'knowledge of the climate', in promoting climate-smart agriculture through discussions and translations of scientific knowledge into practical applications in the business community, in green vocational education, and in education for adults. The Dutch slogan in this respect is 'People make the transition'. In contrast, the UK policy approach is more diffused as the HEIs are conceived as critical actors in the production of scientific knowledge of relevance to CC policy and strategy that are centred on the establishment of an intermediary body (CCC) tasked to link science and policy/society.

Taking as a whole, these two cases attest to the importance of contextual and historical dimensions or path dependencies (Krücken, 2003) in the articulation of a CC/sustainability agenda in the realm of HE. In this respect, the advantages (synergies and convergence) associated with the 'one size policy approach' pursued by the EU and its member states via CoARA should leave space for explorative approaches at the national and local levels given that not only is each national HE system unique in many regards, but the role attributed to different types of HEIs—with differing profiles, competencies, and resources—should also be taken into account. The important question is what is feasible, after all, in CC mitigation in HE, and what roles can be expected from HEIs given that their own emissions are a fraction of the society's total emissions, and that if their role can mainly be persuasive towards other key operators, including the citizenry. Any serious form of societal or international CC mitigation tends to require political and industrial cooperation with HEIs. For example, a recent research innovation method of using satellites to measure the thickness of glaciers or ice sheets is out of the reach of the scientific community as its costs range from tens to hundreds of millions of dollars for launching an imaging satellite (Urban, 2021), not to mention the overall costs of reducing emissions globally.

Future studies, ideally using mixed methods and longitudinal designs, could shed empirical light on the interplay between policy frameworks at the national levels and transnational levels on the one hand, and the strategic responses by HEIs to these frameworks on the other, in Europe and beyond. In addition, there is also a need to both, empirically map out and unpack (analytically) the sets of tensions, dilemmas, trade-offs, and paradoxes that have emerged—at multiple levels—and how different actors attempt to solve them as well as mediate their intended and unintended effects. That said, it is nonetheless clear that as quintessential knowledge production organisation, HEIs all over the world will continue to be a part of policy discussions in the context of global, national, and local efforts to address wicked societal challenges such as CC.

References

Averchenkova, A., Fankhauser, S., & Finnegan, J. J. (2021). The influence of climate change advisory bodies on political debates: evidence from the UK Committee on Climate Change. *Climate Policy, 21*(9), 1218–1233. https://doi.org/10.1080/14693062.2021.1878008

Back to Earth. (2023). Landing real-world impact in research evaluation. Elsevier. Online. Retrieved December 8, 2023, from https://www.elsevier.com/academic-and-government/back-to-earth-download

Benneworth, P., & Charles, D. (2012). University–community engagement in the wider policy environment. In P. Benneworth (Ed.), *University engagement with socially excluded communities* (pp. 223–241). Springer.

Clark, B. R. (1998). *Creating entrepreneurial universities: Organizational pathways of transformation.* Pergamon.

Climate Action Toolkit. (2021). Climate commission for UK higher and further education. Retrieved June 14, 2023, from https://www.fenews.co.uk/skills/climate-action-toolkit-climate-commission-for-uk-higher-andfurther education/

Climate Agreement. (2019). The Hague 28 June 2019. Retrieved June 14, 2023, from https://www.government.nl/documents/reports/2019/06/28/climate-agreement

Climate change adaptation: Policy information. (2022). UK Government. Retrieved June 15, 2023, from https://www.gov.uk/government/publications/climate-change-adaptation-policy-information/climatechangeadaptation-policy-information

Climate Change Committee. (2022). Progress in reducing emissions. 2022 Report to Parliament. Retrieved May 5, 2023, from https://www.theccc.org.uk/publication/2022-progress-report-to-parliament

CoARA. (2023). Coalition for advancing research assessment. Retrieved June 14, 2023, from https://coara.eu/

Confronting the climate emergency. (2021). A commitment from UK universities. Retrieved June 15, 2023, from https://www.universitiesuk.ac.uk/sites/default/files/field/downloads/2021-10/confronting-theclimateemergency.pdf

Cook, J. (2019). Understanding and countering misinformation about climate change. In I. Chiluwa & S. Samoilenko (Eds.), *Handbook of research on deception, fake news, and misinformation online* (pp. 281–306). IGI-Global.

Corazza, L., & Saluto, P. (2021). Universities and multistakeholder engagement for sustainable development: A research and technology perspective. *IEEE Transactions on Engineering Management, 68*(4), 1173–1178.

Cottafava, D., Ascione, G. S., Corazza, L., & Dhir, A. (2022). Sustainable development goals research in higher education institutions: An interdisciplinarity assessment through an entropy-based indicator. *Journal of Business Research, 151*, 138–155.

Cran-McGreehin. (2022). Net zero: why is it necessary? What's the logic behind a net zero emissions target, and who is moving to set one? Energy & Climate Intelligence Unit, Briefings. Retrieved June 15, 2023, from https://eciu.net/analysis/briefings/net-zero/net-zero-why

Cristofoletti, E. C., & Pinheiro, R. (2023). Greening the University?: Assessing the Impact of Sustainability and SDGs in Universities' Values and Strategies. In W. Filho, F. Frankenberger, & U. Tortato (Eds.), *Sustainability in Practice: Addressing Challenges and Creating Opportunities in Latin America* (pp. 111–126). Springer.

De Elera, Á. (2006). The European research area: On the way towards a European scientific community? *European Law Journal, 12*(5), 559–574.

Dixon, E. (2023) 'Existential risk' to research from failure to demonstrate impact. Sector leaders quizzed in Elsevier survey back shift to more holistic methods of evaluating scholarship. *Times Higher Education Newsletter*, November 7, 2023. Research assessment 'fails to encourage impact' | Times Higher Education (THE). Retrieved November 24, 2023.

Djelic, M. L., & Sahlin-Andersson, K. (Eds.). (2006). *Transnational governance: Institutional dynamics of regulation*. Cambridge University Press.

DORA. (2012/2024). San Francisco Declaration on Research Assessment. Home | DORA (sfdora.org). Retrieved December 12, 2024 from https://sfdora.org/about-dora/

Dudley, H., Jordan, A., & Lorenzoni, I. (2022). Advising national climate policy makers: A longitudinal analysis of the UK Climate Change Committee. *Global Environmental Change, 76*, 102589.

Enders, J., & De Boer, H. (2009). *The mission impossible of the European university: Institutional confusion and institutional diversity* (pp. 159–178). Springer.

Fankhauser, S., Averschenkova, A. & Finnegan, J. (2018). 10 years of the UK Climate Change Act. London School of Economics and Political Science.

Retrieved May 5, 2023, from https://www.lse.ac.uk/granthaminstitute/publication/10-years-climate-change-act/

Filho, L., Weissenberger, S., Luetz, J., Sierra, J., Rampasso, I.S., Sharifi, A., Anholon, R., Eustachio, J. & Kovaleva, M. (2023). Towards a greater engagement of universities in addressing climate change challenges. *Scientific Reports 13*, 19030. https://doi.org/10.1038/s41598-023-45866-x

Findler, F., Schönherr, N., Lozano, R., Reider, D., & Martinuzzi, A. (2019). The impacts of higher education institutions on sustainable development: A review and conceptualization. *International Journal of Sustainability in Higher Education, 20*(1), 23–38.

Goddard, J., Hazelkorn, E., & Vallance, P. (Eds.). (2016). *The civic university: The policy and leadership challenges.* Edward Elgar Publishing.

Government of Netherlands. Climate Policy. (2019). Retrieved June 15, 2023, from https://www.government.nl/topics/climatechange/climatepolicy

Grau, F. X., Escrigas, C., Goddard, J., & Ha, B. (2017). Editors' Introduction. In F. X. Grau et al. (Eds.), *Towards a socially responsible higher education institution: Balancing the global with the local.* GUNI Global University Network for Innovation.

Gunasekara, C. (2006). The generative and developmental roles of universities in regional innovation systems. *Science and Public Policy, 33*(2), 137–150.

Hoppe, R., Wesselink, A., & Cairns, R. (2013). Lost in the problem: the role of boundary organisations in the governance of climate change. *Wiley Interdisciplinary Reviews: Climate Change, 4*(4), 283–300.

Kaldeway, D. (2018). The grand challenges discourse: Transforming identity work in science and science policy. *Minerva, 56*(2), 161–182.

Kekäle, J. (2001). *Academic leadership.* Nova Science Publishers.

Kekäle, J., Pietilä, M., & Rintamäki, K. (2023). Broadening the conception of 'what counts'—Researcher assessment reform as a reflection of universities' societal expectations. A paper presented at CHER 35th Annual Conference, Vienna.

Kekäle, J., & Pinheiro, R. (2025). Individual Leadership in Higher Education: Partial Solutions, Unintended Consequences, and the Search for an Alternative. In M. B. Rao, A. Singh, & M. P. Rao (Eds.), *Worldviews and Values in Higher Education: Institutional Governance, Leadership, and Capacity Building* (pp. 127–149). Emerald.

Klofsten, M., Fayolle, A., Guerrero, M., & Mian, S. (2019). The entrepreneurial university as driver for economic growth and social change-Key strategic challenges. Technological Forecasting for Social Change, 141, 149–158. The entrepreneurial university as driver for economic growth and social change—Key strategic challenges—ScienceDirect. Retrieved June 14, 2023.

Krücken, G. (2003). Learning the 'New, New Thing': On the role of path dependency in university structures. *Higher Education, 46*(3), 315–339.

Leal Filho, W., Shiel, C., Paço, A., & Mifsud, M. (2019). Sustainable development goals and sustainability teaching at universities: Falling behind or getting ahead of the pack? *Journal of Cleaner Production, 232*, 285–294.

Leiden Manifesto for Research Metrics. (2015). Leiden manifesto for research Metrics—Home. Retrieved June 14, 2023.

Levin, K., Cashore, B., Bernstein, S., & Auld, G. (2012). Overcoming the tragedy of super wicked problems: Constraining our future selves to ameliorate global climate change. *Policy Sciences, 45*, 123–152.

Mbembe, J. (2016). Decolonizing the university: New directions. *Arts and Humanities in Higher Education, 15*(1), 29–45.

National Inventory Report. (2021). Greenhouse gas emissions in the Netherlands 1990–2019. RIVM report 2021-0007 National Institute for Public Health and the Environment. Retrieved December 8, 2023, from https://www.rivm.nl/bibliotheek/rapporten/2021-0007.pdf

Nowell, L., Dhingra, S., Andrews, K., Gospodinov, J., Liu, C., & Alix Hayden, K. (2020). Grand challenges as educational innovations in higher education: A scoping review of the literature. *Education Research International, 2020*, 1–39.

Our World in Data. (2023). Share of global cumulative CO_2 emissions (ourworldindata.org). Retrieved June 22, 2023.

Pietilä, M., Rintamäki, K., & Kekäle, J. (2023). Broadening the conception of 'what counts'—Example of a narrative CV in a university alliance. Paper presented at the 27th International Conference on Science, Technology and Innovation Indicators (STI 2023).

Policy Paper. (2022). Sustainability and climate change: A strategy for the education and children's services systems. Retrieved June 15, 2023, from https://www.gov.uk/government/publications/sustainability-and-climatechange strategy/sustainability-and-climate-change-a-strategy-for-the-education-and-childrens-servicessystems

Progress Report to Parliament. (2022). The CCC's annual assessment of UK progress in reducing emissions. 2022 Progress Report to Parliament—Climate Change Committee (theccc.org.uk). Retrieved April 27, 2023.

Ramirez, F. O., Byrkjeflot, H., & Pinheiro, R. (2016). Higher education and health organizational fields in the age of "world class" and "best practices". In R. Pinheiro, L. Geschwind, F. Ramirez, & K. Vrangbæk (Eds.), *Towards a comparative institutionalism: Forms, dynamics and logics across the organizational fields of health care and higher education* (pp. 35–57). Emerald.

Recognition & Rewards: Panel discussion. (2021). 20 October 2021. Retrieved March 11, 2022, from https://www.radboudumc.nl/en/news/2021/recognition-and-rewards

Rickards, L., & Pietsch, T. (2022) Climate change is the most important mission for universities of the 21st century. Retrieved June 14, 2023, from https://theconversation.com/climate-change-is-the-most-important-mission-for-universities-of-the-21st-century-139214

Room for Everyone's Talent. (2019). Towards a new balance in the recognition and rewards of academics. Retrieved June 15, 2023, from https://www.nwo.nl/sites/nwo/files/media-files/2019-Recognition-Rewards-Position-Paper_EN.pdf

Satell, G. (2017). The 30 year rule—How long does innovation take? A blog in disruptor league. Retrieved June 19, 2023, from https://www.disruptorleague.com/blog/2017/10/06/the-30-years-rule-how-long-does-innovationtake/

Schwartzman, S., Pinheiro, R., & Pillay, P. (Eds.). (2015). *Higher education in the BRICS countries: investigating the pact between higher education and society*. Springer.

Sivertsen, G., & Rushford, A. (2022). The new European reform of research assessment. R-QUEST Policy Brief no. 7. Retrieved June 14, 2023, from https://www.r-quest.no/wp-content/uploads/2023/02/R-QUEST-Policy-Brief-7.pdf

Sørensen, M. P., Geschwind, L., Kekäle, J., & Pinheiro, R. (2019). *The responsible university: Exploring the Nordic context and beyond*. Springer Nature.

Stensaker, B., & Hermansen, H. (2025). Global, Nordic, or institutional visions? An investigation into how Nordic universities are adapting to the SDGs. *Higher Education*, 89, 293–309.

Tam, J. (2022). Has the government's sustainability and climate change strategy delivered for universities? Insights and analysis, Universities UK. Retrieved June 15, 2023, from https://www.universitiesuk.ac.uk/latest/insights-and-analysis/has-governments-sustainability-and

Tu Delft. (2021). TU Delft Recognition & Rewards Perspective 2021–2024. Online. Retrieved December 8, 2023, from https://filelist.tudelft.nl/TUDelft/Over_TU_Delft/Strategie/Erkennen%20en%20Waarderen%20-0recognition%20and%20reward/TU%20Delft%20Recognition%20and%20Rewards%20perspective%20def

UN. (2022). Knowledge-driven actions: Transforming higher education for global sustainability. United Nations Educational, Scientific and Cultural Organization. Retrieved June 14, 2023, from https://unesdoc.unesco.org/ark:/48223/pf0000380519

UNFCCC. (2015). The Paris Agreement. International treaty on climate change. Retrived December 12, 2024 from https://unfccc.int/process-and-meetings/the-paris-agreement?gad_source=1&gclid=CjwKCAiAjeW6BhBAEiwAdKltMiSjGPeDbdkDBbpYtWN3qTUJaAO0HFrxFLV1pCqt6N2zZRaUSVZPTxoCEg0QAvD_BwE

UN SDG. (2023). Financing and Funding. UN Sustainable Development Group. UNSDG | 2030 Agenda—Financing and Funding. Retrieved June 21, 2023.

United Nations. (2015). Transforming our World: The 2030 Agenda for Sustainable Development. Department of Economic and Social Affairs. Sustainable Development. Retrieved June 14, 2023, from https://sdgs.un.org/publications/transforming-our-world-2030-agenda-sustainable-development-17981

University of Reading. (2021). National Climate Education Action Plan. Retrieved June 14, 2023, from https://www.reading.ac.uk/planet/-/media/project/uor-main/uor-campaign/climate-for-change/climate-education-summit/climateeducationsummit-actionplan.pdf?la=en&hash=70A9DA27CDA84F375723D8C91F45B12F

Urban, K. (2021). How do satellites help us track climate change? American University, Washington, DC. Center for Innovation, and New Technology. Online. How Do Satellites Help Us Track Climate Change? | American University, Washington, DC. Retrieved December 11, 2023.

Veraart, J. A., van Nieuwaal, K., Driessen, P., & Kabat, P. (2014). From climate research to climate compatible development: Experiences and progress in the Netherlands. Springer Link. Regional Environmental Change. Online: Retrieved January 15, 2024, from https://link.springer.com/article/10.1007/s10113-013-0567-7

Waas, T., Verbruggen, A., & Wright, T. (2010). University research for sustainable development: Definition and characteristics explored. *Journal of Cleaner Production, 18*(7), 629–636.

Witte, J. (2021). Higher education, science, and the climate crisis. In S. Marginson, B. Cantwell, D. Platonova, & A. Smolentseva (Eds.), *Assessing the contributions of higher education.* Edward Elgar.

Yin, R. K. (2003). Designing case studies. *Qualitative Research Methods, 5*(14), 359–386.

CHAPTER 3

Reframing, Reacting, or Responding? The Climate Emergency and Higher Education in Ireland

Andrew G. Gibson and Ellen Hazelkorn

Introduction

In a debate on 9 May 2019, the Dáil (the primary house of the Irish parliament) passed a motion declaring a climate and biodiversity emergency, making Ireland the second country in the world to do so (Oireachtas Éireann, 2019a).[1] This declaration followed the discussion of a landmark

[1] In the United Kingdom, the Scottish Parliament declared a climate emergency on 28 April, and the Welsh Senedd the following day 29 April. The devolved jurisdictions of Gibraltar and Jersey made declarations on 2 May and 3 May, respectively. Though a number of bills have been introduced in the United Kingdom Parliament, as of the time of writing, none have passed.

A. G. Gibson (✉)
School of Education, Trinity College Dublin, Dublin, Ireland
e-mail: gibsona@tcd.ie

E. Hazelkorn
Centre for Global Higher Education, Technological University Dublin (TU), Dublin, Ireland
e-mail: ellen.hazelkorn@tudublin.ie

J. Kekäle, R. Pinheiro (eds.), *Higher Education Policy for Tackling Climate Change*, https://doi.org/10.1007/978-3-031-78479-8_3

cross-party *Report of the Joint Committee on Climate Action Climate Change: A Cross-Party Consensus for Action.* In this debate, the responsible member of the government cabinet, the Minister for Communications, Climate Action and Environment, Richard Bruton, commented that it:

> is not the sort of emergency where some individual or Minister is given emergency powers which set aside normal parliamentary procedures or accountability to do something for a short time. This is an emergency where we have to orchestrate our whole community to adopt a different way of dealing with issues. (Oireachtas Éireann, 2019a)

Universities and other higher education institutions (HEIs) in Ireland have been identified in various policies as an important part of addressing this emergency, as they must contribute to the communities in which they are rooted. From a government perspective, the importance of higher education to the economy has recently seen the establishment of a new tertiary education ministry in 2020: the Department of Further and Higher Education, Research, Innovation, and Skills (DFHERIS). This would seem to imply that there is scope for a coordinated approach to the climate emergency across higher education policy and within HEIs in receipt of public funding. For all these indications that the Irish government was beginning to take the climate emergency seriously, Ireland nevertheless remains a "climate laggard" (Torney, 2020).

This chapter addresses the connections across the various policy realms (national, sectoral, and institutional), and what these amount to in practice. The first aspect can be outlining by describing the nature of the policy reaction taken by the Irish state at the national level, the audiences this was addressed to, how this connects with government policy for higher education, and then what is happening within HEIs themselves. While HEIs in Ireland have varying degrees of autonomy, they are not under strict *dirigiste* government control; accordingly, considerable variety across and within universities might be expected. This initial issue of the connections across policy remains descriptive, but is necessary groundwork for the next, more critical interrogation. The second important question which this chapter seeks to answer is to what extent this relatively recent flurry of policy making is a reframing of existing concerns within higher education, or does it amount to a response which rises to the level of seriousness of an 'emergency'?

In order to address these questions, a range of types of policy—understood broadly, with as much focus on "policy making" as "policy mades"

(Gibson & Bengtsen, 2024, p. 8)—are engaged with, in order to describe the approach being taken by the Irish government and Irish higher education. Of course, the scientific activities of individual researchers and research groups are at the heart of how HEIs can respond scientifically to the climate emergency, not to mention the educational activities of those in higher education. Both education and research are the bread and butter of higher education. This chapter, however, considers HEIs themselves as political actors, involved in the crafting of institutional policies related to the climate emergency. The time-period considered in this paper is primarily from 2013 to 2023, in order to establish a snapshot of contemporary Irish higher education and climate change, and to allow consideration of some future directions that may be taken.

Supporting existing research on Ireland's status as a 'climate laggard', this chapter's consciously critical approach to climate policy in higher education allows us to show how generative as well as necessary a critical stance can be. As such, it is worth being aware of the distinction the title for this chapter proposes, namely between policies that reframe existing concerns in the light of climate change, more grudging policy reactions, and finally more proactive policy responses to the climate emergency. This allows for an evaluation of what interventions in policy space set out to do and the degree of ambition they evince. Such a distinction allows us, in this chapter, to show the nature of the reframing of and reaction to climate change made by the Irish state as explaining why Ireland remains a climate laggard (Torney, 2020). It also allows us to point towards the strong calls for a proactive response from Irish society as a counterweight, and finally it explains the mixture of reframing, reacting, and responding we identify on the part of HEIs and those working in higher education.

Methodology and Conceptual Framing: Policy Ontology

The standard, traditional approach to policy is to examine and evaluate a document, considering costs and benefits, assuming that its writing involves (mostly) rational actors whose motivations are transparent to the researcher (Diem et al., 2019). This is effectively the practitioner perspective, "a dominant account of policy which is still regarded as 'common sense' by many policy professionals" (Shore & Wright, 2011, p. 4). It frequently inheres a largely one-dimensional view of power, where conflicts are explicit, and we can study concrete observable behaviour (Lukes,

2005, pp. 16–17). Here the operating assumption is not just that those involved have the political power and will to implement policies, but also that they might improve these policies in response to researchers' findings. Such approaches have variously been termed "technical empirical" (Codd, 1988), "authoritative instrumentalist" (Shore & Wright, 2011), "traditional positivist" (Diem et al., 2019). The questions posed in this chapter, however, accept that there is often a gap between what is set out in policy, and what is occurring in reality. It's just as important to consider what is not put on the agenda, where potential issues are sidelined, and conflict is thereby made latent (Lukes, 2005, pp. 26–27).

The notion of Ireland being a 'laggard' testifies to this fact that while policy appears to make the intended actions of various state elements "legible" (Scott, 1998, p. 80), these may in fact be no more than a kind of state simplification. They may translate issues from the public sphere into acceptable form, reframing them, giving the appearance of action. For this reason, traditional policy research approaches are not appropriate to the nature and scale of the issue of climate change. Recognising at the outset the distance between what is said and what is done is in line with a variety of approaches to policy that can broadly be put under the heading of 'critical policy studies'. Examples of such modes of researching policy abound. One is Carol Bacchi's (2012) 'what's the problem represented to be' (WPR) approach, which recognises that "what one proposes to do about something reveals what one thinks is problematic". Another is Taina Saarinen's work on higher education policy, influenced by critical discourse analysis, where policy actors' "presuppositions seem to 'persuade that', or represent as existing, rather than 'persuade to', or convince the readers that it would be necessary to act" (Saarinen, 2008, p. 723). Gibson and Bengtsen push such analyses of policy to argue that every policy has what they term a 'policy ontology', wherein its "basic ontology or picture of the world is accepted as a given to which we must adapt ourselves" (2024, p. 8). Building off the insights of Bacchi, Gibson and Bengtsen suggest there is a narrowing whereby "the policy ontology of the problem sets out a policy reality system whereby whatever policy captures *is* reality" (2024, p. 15). A final point about critical policy approaches however, is that they are not merely relentlessly suspicious, dismissing documents as "'mere rhetoric' which [have] very little to do with real life policy actions" (Saarinen, 2008, p. 719). There is in this the focus for converting what are identified as issues into "matters of concern" (Latour, 2004).

This also relates to the question of justifying such a critical policy approach, however. It may seem tactless, especially in the context of climate change, to not relentlessly focus on the practical, and simply get on with it as policy practitioners or activists might desire. But there are dangers in this, namely in terms of what Elaine Scarry (2011, pp. 14–15) calls the "seduction against thinking in an emergency". This establishes "a false opposition between thinking and acting; second, [...] a plausible (but in the end, false) opposition between thinking and *rapid* action". As such, there is real need to critically interrogate not just what is said (policy), and what is done (action), but the thoughts and concepts informing both. With specific reference to the area under discussion, the climate emergency, there is also an important critical perspective on the political and economic which challenge the prevailing narrative found in the majority of policies in this area.

Policies that construct the problem as the need to *adapt* to 'climate change'—itself a neutralisation and attempt to naturalise the underlying structural conditions that have given us "global boiling" (Guterres, 2023)—have made a political choice to largely maintain things as they are. Recognising this is the critical perspective that recognises the intentionally narrowed range of options made available to citizens and politicians, what Romain Felli terms "the great adaptation". Such a view recognises that the "climate crisis is a crisis in the capitalist way of organising nature", and as such the "climate crisis allows for the extension of the market's grip" (Felli, 2021, pp. 3, 6). Similar critiques have been offered from within education, and the range of perspectives is captured in the labels for the different approaches, inter alia: environmental education (Hart & Nolan, 1999), education for sustainable development (Disinger, 1990), climate change education (Roussel and Cutter-Mackenzie-Knowles, 2020), climate change education for sustainable development (Anderson, 2012), or climate literacy (McCaffrey & Buhr, 2008). Of these, education for sustainable development (ESD) is the most prevalent, and it aligns with the general acceptance of the importance of 'sustainability' as a contemporary personal and policy virtue, aligning as it does with the UN's 'Sustainable Development Goals'. The attractiveness and general acceptance of 'sustainability' has contributed greatly to ESD as a policy-driven phenomenon (Jickling & Wals, 2008), and the representation of the 'problem' of the climate emergency has been defined for many in policy circles as one of sustainability.

Another important perspective to bring into play when discussing the role of higher education in a time of climate emergency is the danger of "cockpit-ism" (Hajer et al., 2015, p. 1651), which is "the illusion that top-down steering by governments and intergovernmental organisations alone can address global problems". The top-down, cockpit approach saw climate change like a state (Scott, 1998), and led ultimately to the failure of COP15 Climate Summit in Copenhagen in 2009. In contrast, COP21 in Paris in 2015 was arguably a relative success, as it did not assume the top-down perspective as the way to get things done. A lesson from this, as Rebecca Willis sets out, is that science cannot sideline democracy and politics, as "there is no smooth, agreed process by which scientific evidence can be translated into international action" (Willis, 2020, p. 22). An implication of this point of political coordination is that while universities and academic researchers are unable single-handedly to solve global problems, they *can* look after what they have control over, on the one hand, and be an influential voice, as they were for instance during COVID-19, on the other.

The combination of these methodological and conceptual issues inform the approach taken in this chapter. The Irish government's policies as they relate to climate change are not assumed to be the final word, as documents frequently serve the purpose of not denoting what will be done, but merely responding to demands for something to be done. The approach taken is indicative rather than exhaustive, but policy is broadly construed in terms not just of documents and texts, but also changes to governance structures, funding priorities, and so on. As such, national and higher education policies are supplemented by the institutional perspective, to see what universities and other HEIs are doing, and what the academy itself sees as being significant. To draw out something of what 'the academy' itself as a whole considers to be important, like other authors in this volume, we analysed data from the most significant Irish research-funding agencies. For this institutional view, we draw on the valuable research of others in this area, whose results delve into greater detail on the nature of the policy and governance responses that have been and can be taken by individual HEIs.

National Climate Change Policy: Reacting to Crisis

Ireland has been called a "climate laggard" (Torney, 2020) because it has consistently fallen behind its EU commitments, and has been ranked among the lowest performing in the annual Climate Change Performance Index. Despite rising nine spots from 46th in 2022 to 37th place in 2023, out of 59 countries and the EU, Ireland has a poor record due to a medium rating in Renewable Energy and Energy Use, low in Climate Policy and very low in GHG Emissions categories (CCPI, 2023). It continues to fall behind on its commitments (e.g. EPA, 2023). Having said this, since 2017 there has been a flurry of policy and action. The most notable development has been in renewable electricity, building off Ireland's position in the North Atlantic. For a country which was long considered absent of any natural resources save its people, Ireland is now recognised as possessing one of the richest wave and tidal energy climates in the world alongside wind power. Other notable developments include the reports of the Citizens' Assembly on Climate Change 2018 and parliamentary committee established to consider the Assembly's recommendations, and the *Climate Action Plans* published annually between 2019 and 2023.

A possible explanation for the shift in policy zeal is down to a combination of EU and international pressure, public opinion, and the ascendancy of the Green Party to a three-way coalition government in June 2020. In 2018, the EU gave member states 18 months to develop 30-year strategies and submit by January 2020. By the start of 2023, Ireland was one of handful of countries not to have done so however—blaming the delay on the necessity to align the domestic policy with legislation (Boland, 2023). A recent study indicates that the overwhelming majority (70%) of Irish adults are worried about climate change—coming third after housing and healthcare as the most pressing issue. Almost 90% believe human activity is causing the problem, identifying energy and transport usage as the key causes (Timmons & Lunn, 2022). The Green Party had been in government previously, but as Torney argues, the deteriorating political and economic context around 2008–2010 coupled with the "failure of COP15 to produce a meaningful international climate change agreement gave naysayers a political opening to criticise the proposal for a climate law" (Torney, 2017, p. 254). While there is evidence of some policy diffusion via the UK (Torney, 2017, p. 263), arguably the Green Party's recent membership of the three-way ruling coalition coincided with the rising green zeitgeist.

Irish public policy on Climate Change dates to the early to mid-2000s and the publication of two overarching climate change policy documents, Ireland's first (2000) and second (2007) National Climate Change Strategies. However, the country rapidly descended into bankruptcy and had to be rescued by the IMF following the global financial crisis and Great Recession of 2008—ending political and public interest in Climate Change. In 2009, the EU—which had been a partner to Ireland's bailout—enacted legislation underpinning its climate and energy package for

2020 (EC, 2009), with corresponding obligations for Ireland beginning to put pressure on the Irish government.

As the economy emerged from recession in the 2010s, there was more political space to consider and fund initiatives (see Box 3.1 below; note also the changing name of the relevant government departments, with the eventual replacement of 'Energy and Natural Resources' with 'Climate Action and Environment'). Nonetheless, progress has been slow. Its geographic size and small population, along with a political structure favouring a clientelist political culture which benefits close citizen-political relationships, means Ireland thus experiences considerable difficulties moving beyond policy statements and translating 'sound bites' into measurable actions. It is also beholden to the "historical legacy of a predominantly agricultural society [which] gives the farming sector a [outsized] voice in national policymaking" (Torney, 2020, p. 48). For climate change this is no different. It is hindered by an over-reliance on oil and gas for heating, and failure to tackle emissions from agriculture and transport (EPA, 2023; Lynch, 2023) as well as very public opposition from certain sectors to carbon tax and rewetting of bog lands (Leahy, 2023), to name just a few contentious issues.

Box 3.1 Key national policy and legislative actions for climate change and sustainability

- 2000. *National Climate Change Strategy Ireland*. Department of Environment and Local Government.
- 2007. *National Climate Change Strategy 2007–2012*. Department of Environment, Heritage and Local Government.
- 2007. *Delivering a Sustainable Energy Future for Ireland*. Department of Communications, Marine and Natural Resources.
- 2012. *National Climate Change Adaptation Framework*. Department of Environment, Community and Local Government.
- 2015. *Ireland's Transition to a Low Carbon Energy Future, 2015–2030*. Department of Communications, Energy and Natural Resources.

(*continued*)

Box 3.1 (continued)

- 2015. *Climate Action and Low Carbon Development Act.* Irish Statute Book
- 2018. *The Sustainable Development Goals National Implementation Plan 2018–2030.* Department of Communications, Climate Action and Environment.
- 2019. *Climate Action Plan 2019 (CAP19): To Tackle Climate Breakdown.* Government of Ireland.
- 2021. *Climate Action Plan 2021 (CAP21): Securing Our Future.* Government of Ireland.
- 2022. *Climate Action Plan 2023 (CAP23): Changing Ireland for the Better.* Government of Ireland.
- 2023. *Climate Action Plan 2024 (CAP24).* Government of Ireland.

The Citizen's Assembly first convened during 2016–2018 as a form of deliberative democracy, and in line with a trend across several other European countries including Iceland, Estonia, Luxembourg, and Romania (Farrell et al., 2019). While such an assembly is not a way to bypass electoral politics, they are "a practical means by which elected politicians can explore the public mandate for action – developing 'representative claims' which people can support" (Willis, 2020, p. 88). Established by the Irish government in 2016, the process involves a random selection of 100 adult citizens to discuss and propose solutions to a set of contentious public issues. The Dáil passed a resolution establishing the Citizens' Assembly, initially giving it five items to consider: abortion, the challenges and opportunities of an ageing population, fixed-term parliaments, how referenda are held, and how the state can make Ireland a leader in tackling climate change. This "rather eclectic set of topics in part resulted from commitments made by Fine Gael [the primary coalition party] in its manifesto; but in part it reflected international pressures for action in key areas (notably on abortion and climate change)" (Farrell et al., 2019, p. 114). The topic 'How the state can make Ireland a leader in tackling climate change' had four days of meetings in August 2017, with 1185 public submissions, and culminating in the Citizens' Assembly making 13 recommendations for action.

That report's wide range of recommendations for action included putting change considerations at the centre of policy making, energy policy, transport policy, and agriculture and land use policy.[2] Within the first group of questions relating to policy, 97% of members of the Citizens' Assembly recommended that climate change should be at the centre of Irish policy making, while 100% recommended that the state should take a leadership role in addressing climate change across a range of mitigation measures; 80% said they would be willing to pay higher taxes on carbon-intensive activities (Citizens' Assembly, 2018). The radical nature of the recommendations made maps on to more recent opinion polling on what Irish society makes of climate change by the Environmental Protection Agency (EPA, 2022). Here results captured the views of the Irish adult population, across a range of responses from those who are most concerned and motivated about climate change (and with the highest belief in the scientific evidence), to those with the lowest belief in the evidence, and consequently less concerned and motivated. The results were that 36% of respondents were alarmed, 48% concerned, and only 12% cautious of the evidence and the need to do anything, and 3% outright doubtful. Other research qualifies these findings, however, as when presented with a range of issues, respondents place climate third in a list of concerns behind housing and healthcare (two lightning rod crises in contemporary Irish politics), but ahead of the economy or jobs, education and childcare, COVID, crime, taxation, Brexit (the UK's exit from the EU) and the question of a united Ireland, and lastly immigration (Timmons & Lunn, 2022, p. 10). Nevertheless, the results are considerably more radical and demanding than politicians and policy makers have allowed for.

Following the publication of the Citizens' Assembly report, a cross-party parliamentary committee was established to respond to the 17 recommendations. Convening such a cross-party committee signalled political support across the board, with little evidence of climate change being used as a 'wedge-issue' for electoral gain as has been the case in other jurisdictions. That parliamentary committee's report noted "the State's response to climate change has been insufficient" heretofore, and that citizens want the state "to take far more ambitious actions to make Ireland a leader in tackling climate change" (Oireachtas Éireann, 2019b,

[2] Another Citizens' Assembly in 2022 considered biodiversity loss, and published its report the following year (Citizens' Assembly, 2023). This means that to date two of eight Citizens' Assemblies have addressed climate change in some respect.

p. 2). Education was mentioned in this and the above-mentioned report as playing a critical role with respect to 'climate change literacy' and encouraging the uptake of 'solutions'. Significantly it called for increased investment in research, warning it should not be "confined to any one aspect of academia, as contributions can and should be made from all relevant fields of academia (including science, social science and the humanities)" (Oireachtas Éireann, 2019b, p. 33).

Following this, the Irish government's climate action plans have been the defining, national policy statements in response to calls for greater involvement and action by the state.[3] The *Climate Action Plan 2019* (GoI, 2019) was the first major policy statement, accompanied by a set of 183 actions; while education is mentioned at various points, higher education and universities specifically are not. The most recent *Climate Action Plan 2023* (GoI, 2022a) does explicitly point towards a role for higher education, in its annex of actions identifying a number of policy initiatives (many identified in Box 3.2 below), and specific funding activities such as Science Foundation Ireland's Co-centres Programme, which has a focus on climate, sustainable and resilient food systems. This points towards how higher education in Ireland is being represented (in policy at least) as a component of the Irish state's response to the climate emergency.

Higher Education Policy: Reframing or Responding?

Since the late 1960s, Irish higher education has been framed in terms of government-defined 'national priorities', primarily with the view that higher education ought to serve the economy. Such an approach prioritised a 'human capital' perspective, and resulted in the development of a binary system, with the historically academic universities on one side, and a technical/technological sector on the other (Walsh, 2014). The introduction of legislation for the university sector in 1997 enshrined in policy, the social as well as economic roles stating that the university should "support and contribute to the realisation of national economic and social development" (Oireachtas Éireann, 1997). This also resulted in the establishment of various research-funding agencies, such as Science Foundation

[3] These plans accompany *The Sustainable Development Goals National Implementation Plan 2018–2030* as a whole-of-government initiative. Alignment between the two policy agendas can be found on Ireland's Sustainable Development Goals (SDGs) data hub. See https://irelandsdg.geohive.ie.

Ireland (SFI), and what would become the Irish Research Council (IRC). Higher education had long been under the auspices of a (varyingly named) Department of Education, but 2020 saw the establishment of a dedicated ministry, the Department of Further and Higher Education, Research, Innovation and Science (DFHERIS).

On the whole, however, successive Irish governments have been largely content to leave higher education to its own devices. It was the advent of influential factors such as globalisation, the Bologna Process, and economic crisis that led the state to take a more interventionist stance, resulting in "more intensive assertion of authority over HEIs by government departments and agencies, through increased regulation, revised funding instruments and a structural rationalisation agenda" (Walsh, 2018, pp. 489–90; see Gibson & Hazelkorn, 2017 for the dynamics of one such funding instrument). Nevertheless, the specific governing structures of higher education have remained more in line with 'network governance' than strong managerial, dirigiste approaches (Walsh & Loxley, 2015). Irish political culture and structures requires a 'consensus' approach, which "tends to prevent steamrollering policy changes in favour of a slower, deliberative process; zero-sum arguments are less common" (Hazelkorn & Gibson, 2018, p. 265). This consensus appears most usually in the discussion of specific policy proposals, but beyond this, HEIs enact the reality of network governance. As such, though the higher education 'sector' has been active on climate change issues as institutions and as academics/researchers, its collective voice and impact have arguably been weaker. Unless speaking collective on the need for greater funding of higher education, HEIs have been content to follow their own priorities. This further diffuses public demands for action in various areas, as HEIs largely 'go their own way'.

As noted, the main focus of higher education policy over the decades has primarily been as a beacon for mobile capital and talent (Hazelkorn et al., 2015). In recent years, this has turned to focus on skills and employability and sustainable social and economic development with an emphasis on regional balance. Use of the term 'sustainable' refers here to helping ensure Ireland can avoid the severe economic boom-bust legacy that has afflicted the country over the decades rather than simply a green agenda—albeit there is an interdependent relationship between the two. As such, the vocabulary of climate change and that of the more traditional government focus on economic priorities are muddled in a way that serves policy makers' desire to be all things to all audiences.

State Policies for Higher Education and Climate Change

Nevertheless, with specific reference to climate change, as with national policy, strategy and actions for the higher education sector have stepped up in recent years (see Box 3.2 below). These follow on from the priorities set in national policy discussed previously. The most significant of these focus primarily on the education role that HEIs serve in Irish society, focusing specifically on climate education. The *National Strategy on Education for Sustainable Development to 2030 (ESD)* (GoI, 2022b), co-sponsored by the Department of Education, the DFHERIS, and the Department of Children, Equality, Disability, Integration and Youth was first published in 2014 and spanned 2014–2020. It followed the approach aforementioned:

> The National Strategy on Education for Sustainable Development aims to ensure that education contributes to sustainable development by equipping learners with the relevant knowledge (the 'what'), the key dispositions and skills (the 'how') and the values (the 'why') that will motivate and empower them throughout their lives to become informed active citizens who take action for a more sustainable future. (DoES, 2014, p. 3)

It recommended embedding actions for sustainability across teaching, learning, and research. In the first strategy, the emphasis was on introducing more undergraduate and postgraduate programmes relevant to sustainable development with targets being set and monitored by the Higher Education Authority (HEA) towards their achievement (DoES, 2014, pp. 23–24).

Box 3.2 Key higher education policy and initiatives for climate change and sustainability

- 2014. *'Education for Sustainability'—The National Strategy on Education for Sustainable Development in Ireland 2014–2020.* Department of Education and Skills.
- 2018. *ESD to 2030: Second National Strategy on Education for Sustainable Development.* Department of Education.

(*continued*)

Box 3.2 (continued)

- 2020. Energy Efficiency and Decarbonisation Pathfinder Programme 2020 for the Higher Education Sector. SEAI.
- 2021. *Skills for Zero Carbon: The Demand for Renewable Energy, Residential Retrofit and Electric Vehicle Deployment Skills to 2030.* Expert Group on Future Skills Needs.
- 2022. *ESD to 2030: Implementation Plan 2022–2026.* Government of Ireland.
- 2022. *Impact 2030, Ireland's Research and Innovation Strategy.* Department of Further and Higher Education, Research, Innovation and Science.
- 2022. *SFI Co-centres Programme*—Climate, Sustainable and Resilient Food Systems HE. Science Foundation Ireland.
- 2022. *Green Skills for Further Education and Training 2021–2030 Roadmap FE.* Solas.

The second strategy was much more forthright. Accompanied by a 4.5-year action plan for 2022–2026, an interim review is planned for 2026 to evaluate progress and identify any additional actions required for 2030. In recognition of the international context, it sought to emphasise Ireland's alignment with the UNESCO SDGs and especially EU initiatives, e.g. the European Green Deal and Green Transition, Horizon Europe, and the European Strategy for Universities. Notably, it called attention to the European Declaration on Global Education to 2050: the Dublin Declaration which recommitted to "the importance of global education and public engagement" (GoI, 2022, p. 13). As the HEA stated in its submission to the consultation process, the strategy:

> presents valuable opportunities to implement ESD through multiple channels, not only through sustainability-focused programmes and modules, but also through the integration of ESD across curricula; innovative research; student engagement and activism; campus management; community and industry engagement; and international linkages among others. Substantial progress has been made to date, and there are host of innovative and successful initiatives to learn from and build on. (HEA, 2021)

The picture in terms of research is currently somewhat more complicated, as the two existing major research agencies, SFI and IRC, are in a process of being merged to create a new statutory funding agency (Oireachtas Éireann, 2023). As part of this overall aim, a strategy, *Impact 2030, Ireland's Research and Innovation Strategy* purported to be a significant inflection point in the development of research and innovation in Ireland. *Impact 2030* positions research and innovation at the heart of addressing Ireland's societal, economic, and environmental challenges. The claim was that an additional €1bn would be made available to fund research in 2022 (O'Brien, 2022)—though this funding has yet to materialise.

It can be observed, however, that kinds of issues being tackled in the relevant state policies (Box 3.2) remain in the realm of what is deemed politically acceptable (e.g. sustainability, economic innovation), minor redirections of existing capital funding requirements (e.g. retrofitting), or the institutional interests of a newly founded government department (e.g. further education). This indicates a reframing of already existing concerns and imperatives, rather than the response that the scale of the issues confronting us should invite.

The Academy and Institutional Responses

Instead of focusing solely on higher education as a policy sector, however, it is worth paying attention to what higher education itself is doing in the area of climate change. One possible way is to consider what the funding priorities for researchers in higher education are. While there have been attempts by government to explicitly set funding priorities, most clearly a targeted funding mechanism in the shape of the research prioritisation exercise, this was primarily in response to the economic crash of 2008, and such an approach has not become the standard model for how the Irish government approaches higher education research funding (Gibson & Hazelkorn, 2017). Instead, research is funded largely through the two (as yet unmerged) funding agencies, SFI and the IRC. SFI is the largest competitive funding agency, focusing on science and engineering, and the IRC has a focus on both STEM and AHSS. SFI, established in 2000, with a board comprising academic and non-academic members, channels public funding towards research activity in strategic areas and "from the outset was focused on developing economically strategic areas for research and drawing prestigious international scientists into the Irish research system"

(Walsh, 2018, p. 405). Nevertheless, funding proposals are reviewed by international independent reviewers with the knowledge relevant to the proposal. The IRC largely follows the academic-led model as is found in agencies such as the European Research Council, with a board comprised of academic researchers from different disciplines. The focus of the IRC is largely on the 'human capital development' of academic researchers, funding postgraduate/PhD researchers and early career academics (along with more established and senior funding streams) and discipline-specific panels considering funding applications in line with the 'academic excellence'.

Looking at research funded by the two largest research agencies in the past ten years gives some sense of the kinds of topics that researchers who have been successful in receiving funding are interested in.[4] For the IRC, only the number of projects is identifiable, which is to say of the 5503 projects funded by the IRC since 2013, 194 appear to have a clear focus on climate change-related topics. Of the 3235 projects funded by SFI since 2013, 84 appeared to have a direct bearing on climate change. In funding terms, that means approximately €2.16bn awarded by SFI in that period, €77m (3.5%) of that total went to climate change-related topics. While these facts relate only to project titles, this of course is a crude snapshot rather than exhaustive overview. Nevertheless, this may tell us something about the priorities of researchers applying for funding, especially PhD and early career researchers in the case of the IRC (though with the caveat of the possibility of survivorship bias). It certainly tells us about the priorities of those selecting the topics they believe are worth funding (Fig. 3.1).

Considering how HEIs have responded is subject to some of the issues found in the overview of government policy, such as the tension between what is "legible" in policy (Scott, 1998) and what amounts to a substantive commitment. For considering state policy, this is to be found in the distance between discursive commitments and fiscal, Euro commitments to fund these. For HEIs the question is slightly more difficult. HEIs can set out in policy terms their own commitments, such as in university

[4] For both IRC and SFI, we looked at the project titles supplied by researchers the last 10 years of funding (2013–2023). The chosen search terms were [climat*], [environment*], [sustain*], and [warming]. While the databases for the IRC and SFI contain largely comparable data, the SFI data contains more information relating to specific funding amounts, but stops at 2022. IRC data up to 2023 is available. Both resulting datasets were manually cleaned to remove duplicates, as well as false positives such as 'sustained attention', 'digital environment', or 'sustaining Ireland's cultural heritage'.

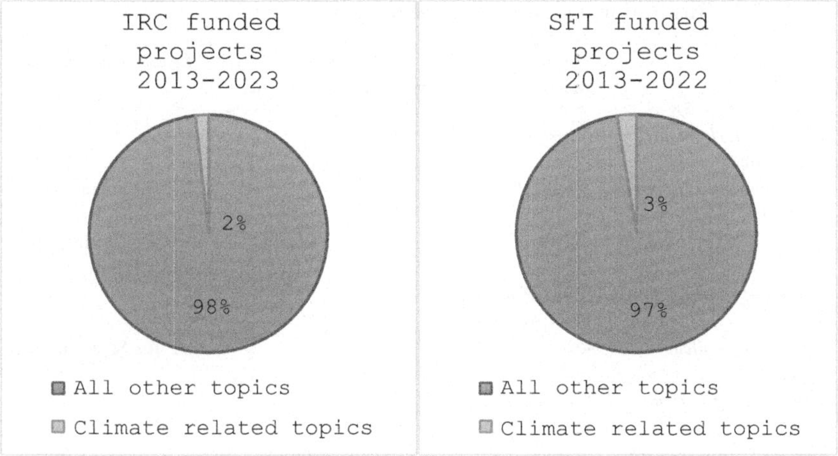

Fig. 3.1 Total of projects funded by IRC and SFI involving (analysis of project names)

strategy statements. One way of considering whether these policy commitments amount to action can be seen in Table 3.1, which sets out those Irish HEIs who have clear and identifiable strategy commitments to climate change—either in their institutional policies or in terms of separate climate/sustainability policies. The question then of whether these policies are backed up at the senior management level, of an identifiable VP position is one way of 'testing' the nature of a HEI's commitment to addressing climate change. There are to be sure limitations with this approach as well, in that larger institutions may have more resources at their disposal, allowing for the appointment of a member of their senior management team. Even with that, it is noteworthy that the only institutions with members of senior management with a climate-relevant brief are all based in the nation's capital, Dublin.

If this approach is too stringent in demanding certain kinds of commitments of HEIs, it is useful then to consider an alternative perspective. In a detailed case study of one university in terms of its governance in responding to climate change, Mary Gallagher-Cooke (2022) sets out the existing routes available to Irish institutions to become a 'generous university' (in our terms one which responds to climate change). These are found across Board- or Governing Authority-level policies, management policies, and

Table 3.1 Institutional policies relating to climate change

	Climate a priority in university strategy	*Separate climate strategy?*	*Leadership position*
Atlantic Technological University	"Sustainability"	–	–
Technological University Dublin	"Planet"	–	VP for Sustainability
Trinity College Dublin	"Civic action"	–	VP for Biodiversity and Climate Action
University College Cork	"Our Place, Our Footprint"	–	–
University College Dublin	–	–	VP for Sustainability
University of Galway	"Sustainability"	–	–
University of Limerick	–	"Sustainable University Strategy 2022–2030"	–

Source: Authors' own, compiled from institutional data

academic policies; the full range captured is indicative of just how much room for manoeuvre universities have. Taking just management policies, Gallagher-Cooke identified the following areas as having the potential to make a significant contribution through, inter alia: financial policies and reporting; energy and emissions policies; green and ethical procurement policies; food policies; water policies; waste and circular policies; commuting and remote/hybrid working policies; corporate travel policies; biodiversity policies and strategy; sustainability in IT and data management; record management; conflict of interest policies, employee induction, probation, promotion; employee learning, training, development, and conference allowance; employee volunteering; employee wellbeing, physical and mental health policies; employee secondment and sabbatical policies. Looking to the future, other resources for change are identified in the autonomous governance of institutions, in terms of either existing policy instruments to amend or new ones to be introduced (Gallagher-Cooke, 2023, p. 13). Many of these are available to HEIs, and do not require mandated action by the state, just the will to make a transformation beyond business as usual.

Future Directions: From Policy Announcement to Policy Enactment

The distance between what is said and what is done in policy may be chalked down to the 'expressive function' (Sunstein, 1996) of creating or recreating norms.[5] Given the stakes, however, this interpretation seems unnecessarily charitable. Rebecca Willis (2020, p. 70) calls for being wary of the "feelgood fallacy" in policy responses to the climate emergency, the "danger that people and organizations channel their energy into the easier sell, encouraging the positives rather than fighting the negatives". Willis saw Ireland's Citizens' Assemblies as avoiding just this trap by recommending "higher taxes on carbon-intensive activities" (p. 90). A more critical perspective would be to recognise that while such *calls* for action avoid that trap, the actual proof needs implementation. The critical policy perspective taken in this chapter is justified, since the show of "'new politics', which seemed to provide the basis for progress on climate issues at the elite level, was stifled by a government that wanted to proceed at a slower pace"; while there may be some evidence for moving beyond such policy stifling in recent years, nevertheless this is a "cautionary tale for supporters of more progressive environmental policy" (Fitzgerald et al., 2021, p. 47). The Irish state needs to move beyond its traditional mode of policy announcement, towards requiring policy enactment and effectiveness.

As of today, however, the claims made by government have not amounted to much for higher education. Indeed, if only looking at the €307m annual shortfall in core funding of universities which the Irish government itself identified, the Minister for Further and Higher Education Simon Harris announced a mere €60m increase following the release of the 2024 Government budget in October 2023, with the majority of funding aimed at hard-pressed students in the shape of additional supports (O'Brien, 2022). There continues to be little evidence of follow-through on the grand announcements for Irish higher education beyond the claims made in policy.

As such it may make sense for those in higher education to retain a considerable degree of scepticism regarding 'cockpit-ism'. The specific role played by the Irish state in 'stifling policy' points towards other actors. Indeed "[d]epending on the extent to which certain institutions can operate independently of the executive, these could include central banks,

[5] We're thankful to Mary Gallagher-Cooke for this reference.

government agencies, scientific or advisory bodies, and (in less liberal democratic contexts) the courts" (Fitzgerald et al., 2021, p. 48). This group could and should include HEIs and the academy more widely. The fact is that unlike in many other jurisdictions, Irish higher education is a trusted actor to speak on scientific issues, as COVID-19 illustrated (Gibson et al., 2023, p. 213). Recognising this, discourses of climate delay (Lamb et al., 2020) are less likely to result in politically distracting debates.

Another aspect of the limits of the existing approaches is in terms of the relentless focus on 'sustainability' as the most politically acceptable approach, given that it maintains that only relatively minor adjustments are necessary. While there is not much use calling for 'more radical' approaches simply as a kind of reflex, what is most important is that they are the correct and proportionate approaches. It may be that 'sustainability' served a purpose in getting some of the issues on the wider agenda, but it is past time to move further. To date, the problem has been represented to be an issue of adjustment, which attempts to smooth over the contradictions in a political-economic model that has not yet confronted the extreme cost of the externalities it inflicts on our world. In terms of the one-dimensional approach to power Lukes (2005) identifies, accepting the basic premise that sustainability and our current politico-economic structures are fit for purpose no longer follows—and in Ireland there is sufficient evidence to show this is the case. In their detailed critique of education for sustainable development, Selby and Kagawa (2010, pp. 42–44) set out a contrasting vision of climate change education (CCE) that acknowledges that "climate crisis that is already upon us" and is likely to deepen further, and so among other things CCE "needs to offer alternative conceptions of the 'good life', combat consumerism, and help learners explore and experience alternatives to a growth economy". They also call for 'a localisation of focus', which is where HEIs might find a role.

This has implications for how the university is organised, and the kind of research given attention, calling into question the usual focus on "natural – and especially climate – sciences provided the framing for the problem", with the social sciences only then being "called on to imagine what the response should be" (Felli, 2021, p. 45)—primarily within a pregiven framework. While "most proponents of ESD seem to have found a space where they feel they can more or less shrug off the need for deep critical reflection", to reverse the dominant "instrumental and exploitative 'nature as resource' philosophy, CCE needs to embrace a philosophy of intimacy with, and embeddedness in, nature not least through the cultivation of the poetic" (Selby & Kagawa, 2010, pp. 39, 44). Through the sustainability perspective it was decided what the 'problem' should be, and what the policy ontology was, with this framing by definition excluding an engagement with alternatives through the social sciences and humanities, thus "ignoring society's capacities for innovation and imagination – and even the indeterminacy of the future itself" (Felli, 2021, p. 5). Pushing beyond sustainability will allow not just for a clear-eyed confrontation of the issue, but also may allow a variety of research and education perspectives to enter where previously they had been sidelined.

There is scope within higher education to decide its own approach, and to set its own priorities, without waiting to be directed or instructed. HEIs have a range of policy options at their disposal, just as the academy has long had ways of steering the direction of future research through funding calls and the establishment of research programmes. Not everything in the university can or will be transformed entirely to address the climate crisis, but unnecessary barriers can at the very least be removed. Academic forms of association and organisations (networks and associations) are powerful social modes of organisation available to higher education and the academy, and which are an independent route to political and policy influence. Society is not the state, and the university needs to remember that whatever its reliance on the latter for (ever decreasing) funding, it is to the former that its responsibilities lie. As yet, Irish higher education (either from the state policy direction or from within HEIs and the academy itself) has seemed content to reframe its existing concerns and priorities, incremental shifts in focus, and the development of a comprehensive suite of policies. As ever in the Irish context, however, implementation lags behind. The need for action *and* the need for careful thinking about the kinds of action that should be taken should be combined, and Irish higher education is well placed to do this difficult work.

The variety of ways in which relevant topics are discussed—'sustainability', 'decarbonisation', 'the green campus', 'climate and biodiversity', 'sustainable development goals', 'national challenges'—is indicative of a complex picture of existing and new concerns. For future research in this area, it may prove useful to proceed with this chapter's evaluative typology—reframing, reacting, responding—in order to focus on the most effective and transformational policies which are possible. It is perhaps inevitable that there will be a certain amount of 'reframing' of existing pressures in higher education, for instance in speaking of energy concerns which have as much to do with efficiency or value for money in public spending in terms of a 'green transition', or including the word 'climate' in abstracts in the hope of attracting research funding for projects with only an incidental relation to global boiling. HEIs are not immune from such forms of greenwashing in this respect. The nature of the policy structure in Ireland, as in other countries, also inevitably implies that HEIs will frequently have to 'react' to directives and legislation, and develop institutional ways to enact these policies. These are probably on the whole beneficial, but only if they are not understood as the last word in this area. It is the final type of policy action we identify however, namely proactive responses, where there is most cause for attention—and hope.

Indeed, in policy and practical terms, the clear indication of an appetite for a strong response on the part of Irish civil society (e.g. via the Citizens' Assemblies) should be interpreted by those involved in the governance of HEIs as a green light to push even further with the positive, proactive changes they are considering. Higher education is well understood as having a significant role as a producer, repository, and translator of knowledge for society. Perhaps given less consideration, however, is the role of higher education as an ontological exemplar, creating a space for experimentation and openness and modelling new ways of being. Both research and policy should continue to focus on doing justice to the public demand for a response, thinking in the climate emergency, and exploring ways to make change possible.

References

Anderson, A. (2012). Climate change education for mitigation and adaptation. *Journal of Education for Sustainable Development*, *6*(2), 191–206. https://doi.org/10.1177/0973408212475199

Bacchi, C. (2012). Introducing the 'What's the Problem Represented to be?' approach. In A. Bletsas & C. Beasley (Eds.), *Engaging with Carol Bacchi: Strategic Interventions and Exchanges* (pp. 21–24). University of Adelaide Press. https://doi.org/10.1017/UPO9780987171856.003

Boland, L. (2023, April 28). Ireland has finally released its 30-year climate plan – here's what it says. *The Journal*. https://www.thejournal.ie/long-term-climate-strategy-6055747-Apr2023/

CCPI. (2023). Climate change performance index, Ireland. Retrieved September 16, 2023, from https://ccpi.org/country/irl/

Citizens' Assembly. (2018). *Third Report and Recommendations of the Citizens' Assembly. How the state can make Ireland a leader in tackling climate change.* Government Publications. https://citizensassembly.ie/wp-content/uploads/2023/02/ThirdReportClimate-Change-Report-Final.pdf

Citizens' Assembly. (2023). *Report of the citizen's assembly on biodiversity loss*. Government Publications. https://citizensassembly.ie/wp-content/uploads/ReportonBiodiversityLoss.pdf

Codd, J. A. (1988). The construction and deconstruction of educational policy documents. *Journal of Education Policy, 3*(3), 235–247. https://doi.org/10.1080/0268093880030303

Diem, S., Young, M. D., & Sampson, C. (2019). Where critical policy meets the politics of education: An introduction. *Educational Policy, 33*(1), 3–15. https://doi.org/10.1177/0895904818807317

Disinger, J. F. (1990). Environmental education for sustainable development? *The Journal of Environmental Education, 21*(4), 3–6.

DoES. (2014). "Education for Sustainability". The national strategy on education for sustainable development in Ireland 2014–2020 (Issue July). https://www.gov.ie/en/publication/02952d-national-strategy-on-education-for-sustainable-development-in-irelan/

EC. (2009). *2020 climate & energy package*. European Commission. https://climate.ec.europa.eu/eu-action/climate-strategies-targets/2020-climate-energy-package_en

EPA. (2022). *Climate change's four Irelands: An audience segmentation analysis*. Environmental Protection Agency.

EPA. (2023). *Ireland's greenhouse gas emissions projections 2022–2040*. Environmental Protection Agency.

Farrell, D. M., Suiter, J., & Harris, C. (2019). 'Systematizing' constitutional deliberation: The 2016–18 citizens' assembly in Ireland. *Irish Political Studies, 34*(1), 113–123. https://doi.org/10.1080/07907184.2018.1534832

Felli, R. (2021). *The great adaptation*. Verso.

Fitzgerald, L. M., Tobin, P., Burns, C., & Eckersley, P. (2021). The 'Stifling' of new climate politics in Ireland. *Politics and Governance, 9*(2), 41–50.

Gallagher-Cooke, M. (2022). *Towards a Generous University: UCD's pathways to embodiment of sustainability*. MSc thesis, School of Politics and International Relations, University College Dublin.

Gallagher-Cooke, M. (2023). Higher education institution policies in the climate and ecological emergency. *PublicPolicy.ie*. https://publicpolicy.ie/environment/higher-education-institution-policies-in-the-climate-and-ecological-emergency/

Gibson, A. G., & Bengtsen, S. S. (2024). World humanities – Towards an ontology of policy. *Arts and Humanities in Higher Education, 23*(1), 3–22. https://doi.org/10.1177/14740222231189806

Gibson, A. G., & Hazelkorn, E. (2017). Arts and humanities research, redefining public benefit, and research prioritization in Ireland. *Research Evaluation, 26*(3), 199–210. https://doi.org/10.1093/reseval/rvx012

Gibson, A. G., Hazelkorn, E., & Walsh, J. (2023). Pandemic study: Ireland case study. In A. Veiga, & T. Seidenschnur (Eds.), *The role of European universities in an age of pandemic*. Working papers on university reform, no. 37. Centre for Higher Education Futures. https://dpu.au.dk/fileadmin/edu/Forskning/CHEF/Publications/CHEF_Working_paper_Pandemic_study_2023_Final_Version.pdf

Government of Ireland. (2019). *Climate action plan 2019: To tackle climate breakdown*. Government Publications. https://assets.gov.ie/25419/c97cdecddf8c49ab976e773d4e11e515.pdf

Government of Ireland. (2022a). *Climate action plan 2023 (CAP23): Changing Ireland for the better*. Government Publications. https://www.gov.ie/en/publication/7bd8c-climate-action-plan-2023

Government of Ireland. (2022b). *ESD to 2030: Second national strategy on education for sustainable development*. Government Publications. https://www.gov.ie/pdf/?file=https://assets.gov.ie/228330/c69895a6-88f0-4132-b6d1-9085a9c31996.pdf#page=null

Guterres, A. (2023). Secretary-General's press conference – On climate. July 27. United Nations. https://www.un.org/sg/en/content/sg/press-encounter/2023-07-27/secretary-generals-press-conference-climate

Hajer, M., Nilsson, M., Raworth, K., Bakker, P., Berkhout, F., de Boer, Y., Rockström, J., Ludwig, K., & Kok, M. (2015). Beyond cockpitism: Four insights to enhance the transformative potential of the sustainable development goals. *Sustainability, 7*(2), 1651–1660. https://doi.org/10.3390/su7021651

Hart, P., & Nolan, K. (1999). A critical analysis of research in environmental education. *Studies in Science Education, 34*(1), 1–69. https://doi.org/10.1080/03057269908560148

Hazelkorn, E., & Gibson, A. G. (2018). Public goods and public policy: What is public good, and who and what decides? *Higher Education, 78*(2), 257–271. https://doi.org/10.1007/s10734-018-0341-3

Hazelkorn, E., Gibson, A. G., & Harkin, S. (2015). From massification to globalisation: Reflections on the transformation of Irish higher education. In K. Rafter & M. O'Brien (Eds.), *The state in transition – Essays in honour of John Horgan* (pp. 235–260). New Island Books.

HEA. (2021). *National strategy on education for sustainable development*. Submission of the Higher Education Authority (HEA) to the Department of Further and Higher Education, Research, Innovation and Science (DFHERIS). Higher Education Authority. https://www.into.ie/app/uploads/2021/06/20210629__INTO-Response-to-Education-for-Sustainable-Development-ESD_final.pdf

Jickling, B., & Wals, A. (2008). Globalization and environmental education: Looking beyond sustainable development. *Journal of Curriculum Studies, 40*(1), 1–21.

Lamb, W. F., Mattioli, G., Levi, S., Roberts, J. T., Capstick, S., Creutzig, F., Minx, J. C., Müller-Hansen, F., Culhane, T., & Steinberger, J. K. (2020). Discourses of climate delay. *Global Sustainability, 3*, e17. https://doi.org/10.1017/sus.2020.13

Latour, B. (2004). Why has critique run out of steam? From matters of fact to matters of Concern. *Critical Inquiry, 30*(2), 225–248. https://doi.org/10.2307/1344358

Leahy, P. (2023, May 23). Varadkar sounds alarm on rewetting proposals back by Green Party. *The Irish Times.*

Lukes, S. (2005). *Power: A radical view*. Palgrave Macmillan.

Lynch, M. (2023, June 25). Don't be distracted by data centres. Here's the real reason we're missing our climate targets. *The Irish Times.*

McCaffrey, M. S., & Buhr, S. M. (2008). Clarifying climate confusion: Addressing systemic holes, cognitive gaps, and misconceptions through climate literacy. *Physical Geography, 29*(6), 512–528. https://doi.org/10.2747/0272-3646.29.6.512

O'Brien, C. (2022, May 18). Science foundation and research council to merge in major shake-up. *The Irish Times.*

Oireachtas Éireann. (1997). *Universities act 1997.* Stationery Office.

Oireachtas Éireann. (2019a). *Report entitled 'Climate Change: A Cross-Party Consensus on Climate Action': Motion*, May 19. Dáil Éireann debate – Thursday, May 9, Vol. 982 No. 5 Houses of the Oireachtas. https://www.oireachtas.ie/en/debates/debate/dail/2019-05-09/32/

Oireachtas Éireann. (2019b). *Report of the joint committee on climate action climate change: A cross-party consensus for action.* March 27. Houses of the Oireachtas.

Oireachtas Éireann. (2023). Joint Committee on Education, Further and Higher Education, Research, Innovation and Science Report on Pre Legislative Scrutiny of the Research and Innovation Bill, 2023. Houses of the Oireachtas. https://data.oireachtas.ie/ie/oireachtas/committee/dail/33/joint_committee_on_education_further_and_higher_education_research_innovation_and_science/reports/2023/2023-07-13_report-on-pre-legislative-scrutiny-of-the-research-and-innovation-bill-2023_en.pdf

Rousell, D., & Amy Cutter-Mackenzie-Knowles. (2020). A systematic review of climate change education: Giving children and young people a 'voice' and a 'hand' in redressing climate change. *Children's Geographies, 18*(2), 191–208. https://doi.org/10.1080/14733285.2019.1614532

Saarinen, T. (2008). Position of text and discourse analysis in higher education policy research. *Studies in Higher Education, 33*(6), 719–728. https://doi.org/10.1080/03075070802457090

Scarry, E. (2011). *Thinking in an Emergency.* W.W. Norton & Co.

Scott, J. C. (1998). *Seeing like a state.* Yale University Press.

Selby, D., & Kagawa, F. (2010). Runaway climate change as challenge to the 'Closing Circle' of education for sustainable development. *Journal of Education for Sustainable Development, 4*(1), 37–50. https://doi.org/10.1177/097340820900400111

Shore, C., & Wright, S. (2011). Conceptualising policy: Technologies of governance and the politics of visibility. In C. Shore, S. Wright, & D. Però (Eds.), *Policy worlds: Anthropology and the analysis of contemporary power* (pp. 1–26). Berghahn Books.

Sunstein, C. (1996). On the expressive function of law. *University of Pennsylvania Law Review, 144*(5), 2021–2053.

Timmons, S., & Lunn, P. (2022). Public understanding of climate change and support for mitigation. *ESRI Research Series, 135*. https://www.esri.ie/publications/public-understanding-of-climate-change-and-support-for-mitigation

Torney, D. (2017). If at first you don't succeed: The development of climate change legislation in Ireland. *Irish Political Studies, 32*(2), 247–267. https://doi.org/10.1080/07907184.2017.1299134

Torney, D. (2020). Ireland's policy response to climate change: An historical overview. In D. Robbins, D. Torney, & P. Brereton (Eds.), *Ireland and the climate crisis* (pp. 37–54). Palgrave. https://doi.org/10.1007/978-3-031-07753-1_7

Walsh, J. (2014). The transformation of higher education in Ireland, 1945–80. In A. Loxley, A. Seery, & J. Walsh (Eds.), *Higher education in Ireland: Practices, policies and possibilities* (pp. 5–32). Palgrave Macmillan.

Walsh, J. (2018). *Higher education in Ireland, 1922–2016*. Palgrave Macmillan.

Walsh, J., & Loxley, A. (2015). The Hunt Report and higher education policy in the Republic of Ireland: 'An international solution to an Irish problem?' *Studies in Higher Education, 40*(6), 1128–1145. https://doi.org/10.1080/03075079.2014.881350

Willis, R. (2020). *Too hot to handle? The democratic challenge of climate change*. Bristol University Press.

CHAPTER 4

Government R&D Funding Policy for Academic Research in Italy: Are There Incentives for Climate Change Solutions?

Emanuela Reale and *Andrea Orazio Spinello*

Introduction

Climate change is a complex and pressing issue of our time, with far-reaching implications that extend beyond environmental concerns to encompass the economic and social spheres. The Sixth Assessment Report from the Intergovernmental Panel on Climate Change emphasizes the urgent need for immediate action to address climate risks since the window of time to limit global temperature rise is rapidly narrowing and the effects of human activities are causing more losses and damage to nature and people than previously estimated (IPCC, 2022). Rethinking the way we live, produce, and consume requires a collaborative effort at various levels of society, with scientific institutions called upon to play a leading role. In this regard, public investment in R&D on climate change can have

E. Reale (✉) • A. O. Spinello
CNR-IRCrES—Research Institute on Sustainable Economic Growth, National Research Council of Italy, Rome, Italy
e-mail: emanuela.reale@cnr.it; andreaorazio.spinello@cnr.it

J. Kekäle, R. Pinheiro (eds.), *Higher Education Policy for Tackling Climate Change*, https://doi.org/10.1007/978-3-031-78479-8_4

a significant impact on many activities, such as the development of new energy conversion technologies and the promotion of sustainable practices; additionally, data from climate change research can inform policy decisions and promote global climate action. Research policies promoted at the state level may either facilitate or hinder the contribution of scientific activities in addressing climate change by providing regulatory frameworks, offering adequate research funding, or setting climate-related goals (Witte, 2023). As key actors in the generation of knowledge, higher education institutions (HEIs)[1] may respond to the incentives provided by the available funding instruments by deciding to prioritize this topic in their research agenda and by mobilizing efforts to access such funding.

The aim of this chapter is to shed light on the structure and the effects of the government R&D policies in Italy for sustaining or increasing academic research on topics related to climate change, examining one important funding instrument used by the government to steer the direction of research activities. The research interest takes on special significance when applied to areas that are highly impacted and vulnerable to climate change risks, such as the Mediterranean region (Lionello & Scarascia, 2018). For this reason, this chapter focuses on Italy, which, due to its geography and configuration, presents a high risk of extreme events that negatively impact ecosystems and human activities, such as heat waves, droughts, and floods (Castellari et al., 2014; Spano et al., 2020).

Demand for Italian academic research to more effectively address climate solutions became more pressing as the public debate on the effects of climate change intensified in the 2000s, but dedicated R&D policies in Italy appear to have remained underdeveloped until recently (Macchiati, 2009). Still, in 2022, a group of eminent Italian scholars from different scientific fields wrote an open letter to policymakers in which they "*urge political parties to consider the fight against climate change as the necessary basis for achieving equitable and sustainable development in the years to come*". Moreover, they took a clear stance, declaring that "*as climate scientists we are willing to contribute to concrete, practicable and effective science-based solutions and actions, but we strongly urge politicians to consider climate crisis a priority, because it undermines our future*" (Barbante et al., 2022). About one year later, in May 2023, the uproar caused by the devastating flooding in Emilia-Romagna drew the country's attention to the

[1] In this chapter, the words "HEIs" and "universities" will be used to refer to research universities.

issue of climate change and encouraged the involvement of researchers to address the problem.

This chapter deals with the following research questions: (i) How do government funding instruments create incentives to steer academic research toward exploring climate solutions? (ii) How do research groups at HEIs react to these incentives? It is worth noting that the latter is particularly interesting due to its limited exploration in the existing literature. We address the research questions using a conceptual framework developed to comprehend the scheme of opportunities *provided* by government funding instruments (R&D programs) and how beneficiaries *perceive* and *respond* to these opportunities (Reale et al., 2014).

The chapter is structured as follows. Section "Conceptual Framework" introduces the conceptual framework that forms the basis for the analysis. Section "Method and Data" outlines the methodology used in the study, along with the data sources. Section "Italy's Political Approach to Climate Change" provides an overview of Italy's climate change policies, highlighting the most significant milestones in the country's efforts to tackle this issue. Section "Italian R&D Policies Addressing Climate Change" is devoted to describing the Italian R&D policies related to the topic. Section "Climate Change Projects Funded Under the PRIN Program" presents the findings of the analysis of how academic research projects in Italy have responded to the incentives from public research policies to address climate change research. Section "Doctoral Theses on Climate Change" investigates the propensity of Italian universities to focus their research on climate change, regardless of the government incentives offered. Finally, Section "Discussion and Conclusions" draws conclusions based on the study's findings. The study highlights Italy's lack of dedicated funding for climate change research; however, this limitation has not deterred universities from conducting research on the topic, demonstrating a certain degree of potential for climate change research.

Conceptual Framework

Public research funding reforms implemented in many countries over the past few decades have sought to improve strategic planning capacity and devote more attention to the need to address topics of social relevance (Foray et al., 2012; Mazzucato, 2018). Government funding for mission-oriented research linked to output and performance criteria has increased. The growing number of competitive funding schemes in the portfolios of research funding organizations has been one method of nudging scientific research toward social and technological priorities, combining the promotion of more selective methods for the distribution of funding on the one hand with the impetus toward efficient use of public resources on the other (Lepori et al., 2007). Particularly, the topic of climate change gained momentum in the realm of European countries' R&D policies from the design of the Societal Grand Challenges within the European Framework Programmes, which gave the national governments a set of global priorities to be addressed through dedicated funding instruments (Kuhlmann & Rip, 2018).

Understanding how R&D government policies are implemented by national research funding organizations and the extent to which government funding can effectively address problems of social relevance defined at policy levels remains a major challenge (Lepori & Reale, 2019; Schneider et al., 2019; Spinello et al., 2021). Several methods have been developed to investigate whether the targets and outcomes of research funding initiatives are achieved, usually with an emphasis on the ability of funding schemes to meet research policy objectives with an impact on societal transformations (Guthrie et al., 2013; Pedersen et al., 2020). The shift in the focus from the functioning of the programs to the relationship and dynamics between actors, such as research funding organizations and research groups, was a significant development for evaluation purposes (Lepori & Reale, 2012). One proposed approach is based on considering the opportunities offered by the programs and how they are perceived by the beneficiaries (Reale et al., 2014). In this regard, research funding instruments represent a set of opportunities *intended* by policymakers and then *provided* to potential beneficiaries through the program design; the opportunities are then *perceived* and *mobilized* differently by the beneficiaries based on the combination of the researchers' own research priorities and the institutions' strategic goals (Reale et al., 2014).

Within this framework, the intended opportunities are linked to policy objectives and rationales, which can be derived by understanding the missions and objectives of funding mechanisms. Whether or not the intended opportunities have been provided by a program depends on the way in which they are implemented through different practices, including selection and funding practices. The beneficiaries perceive each research funding program as a one-of-a-kind opportunity based on (i) how the program is presented and (ii) the importance they place on the opportunity. Opportunities mobilized by beneficiaries consist of the real benefits that motivate beneficiaries to seize funding opportunities. Inconsistencies between *intended* and *provided* opportunities are evidence of possible shortcomings in policy action in program design; inconsistencies between *provided* and *perceived* opportunities shed light on problems with program implementation. Differences affecting *perceived* and *mobilized* opportunities provide evidence of different strategies for participation in competitive programs.

We apply a conceptual framework based on the "opportunity approach" to the Italian R&D funding system for academic research to learn more about the effectiveness of the incentives for climate change research. Italy appears to be a very active country in terms of publications on climate change. According to Romero-Perdomo et al. (2022), it had a very high production rate in terms of works on the integration of topics related to climate change and the circular economy between 2010 and 2020, with the highest number of single-country publications, followed by the United Kingdom and Spain. Furthermore, a bibliometric analysis by Overland and Sovacool (2020) found more than 400 projects from 1990 to 2018 supporting social science climate research. We are interested in studying how government public funding policies have provided incentives for academic research on climate change in recent years, as well as how academic researchers have perceived and mobilized those potential incentives.

Method and Data

This chapter uses a variety of empirical evidence derived from (i) a documentary analysis of Italian climate change policies, (ii) literature on R&D funding instruments in Italy, and (iii) data from two datasets developed within the European Research Infrastructure for Science and Innovation Studies (RISIS)[2]—EFIL, the European dataset of public R&D funding instruments, which contains data to investigate public R&D funding in Europe at the level of project-based funding instruments and research funding organizations (Reale et al., 2023), and the NATPRO module of EUPRO database, which stores data on R&D projects funded by funding schemes managed by national funding agencies (Zahradnik et al., 2021)—integrated with official data from the research general directorate of the Italian Ministry of University and Research (MUR). Finally, to further evaluate the propensity of HEIs to address climate change topics

[2] Website: https://www.risis2.eu/

regardless of government incentives, we also analyzed data on the doctoral theses defended at Italian universities.

The picture that emerges from the analysis of the literature on national policies and the portfolio of funding instruments sheds light on the incentives offered by government policies to address issues related to climate change. The control between the incentives and the mobilization of academic research activities was carried out using data on projects funded by the major funding instrument specifically devoted to academic research in Italy: the program for *Research Projects of National Relevance* (Progetti di Rilevante Interesse Nazionale, PRIN). We analyzed how HEIs have responded to PRIN calls with the submission of research projects on climate change by combining EFIL/NATPRO data integrated with data from the PRIN calls for proposals' portal of the research general directorate of the MUR.[3] We analyzed funded projects within the PRIN calls from 2009 to 2020, collecting data on the following variables: (i) title and abstract of the approved projects, (ii) budget received within the total budget of the call, (iii) universities participating in the projects, and (iv) the scientific sector in which the projects are classified by the MUR, according to the European Research Council (ERC) classification (ERC domain).

In order to identify funded projects related to climate change research, we used a method developed by Overland and Sovacool (2020). This method entails searching for words taken from a long list of 89 keywords related to climate change topics. We used these keywords in English and Italian, combining them with the Boolean operator "OR" and applying them to the titles and abstracts of funded projects. We supplemented the list with a string of another ten pertinent keywords which we had selected as relevant.[4] A shorter string with nine keywords closely related to climate change, such as "climate change" and "global warming", was used to create a smaller boundary for the analysis of the ranges. Universities participating in the selected projects were geolocated and classified as large, medium, or small based on the number of students enrolled in the

[3] Website: https://prin.mur.gov.it/

[4] The search string was integrated with the following keywords: "climatic conditions" OR "climate resilient" OR "CO2 markets" OR "solar cells" OR "CO2 sequestration" OR "solar cooling" OR "energy performance" OR "solar devices" OR "green diesel" OR "environmental changes".

2017/2018 academic year, as reported in the MUR National Student Registry.[5]

To complement the analysis, we also examined the propensity of Italian universities to self-select climate change research as a topic of interest, regardless of the incentives intended and provided by the available funding schemes. Therefore, we used the number of doctoral theses defended at Italian universities as a proxy for the propensity to conduct new research on these topics. We gathered information from the repository of doctoral dissertations at the National Central Library of Rome (Biblioteca Nazionale Centrale di Roma, BNCR), using a metadata collection updated as of November 2021.[6] We extracted the raw information for 2010–2019, which we codified and cleaned. This allowed us to generate standardized data on (i) university affiliation, (ii) year of graduation, coinciding with the year of submission of the thesis at the library, (iii) thesis title, and (iv) scientific field of the doctoral dissertation, which we categorized under the ERC domain. Also, in the case of doctoral theses, we used the selection method based on searching for climate change-related keywords. Lastly, the universities were geolocated and classified as large, medium, or small according to the number of enrolled students (see footnote 5).

For the PRIN projects submitted in two specific calls—2017 and 2020—we were able to conduct the keyword search only on the title of the projects, with the risk of underestimating the projects to be included. As to the PhD theses, the analysis was only possible for the titles and for the theses submitted to the national libraries, for which we calculated the percentage of missing data to be between 22% and 28% starting from 2016, according to MUR data on research doctorates in Italy. We checked all PRIN projects and doctoral theses selected with the automated keyword search to verify their effective relevance to climate change issues using the available data and excluded a few on this basis.

This chapter has some limitations, the most important being that the analyses do not consider important aspects of Italian research on climate change funded by the government, such as its quality, the impact of the results of the projects, and the impact of the doctoral theses. Another

[5] Large universities are the top 20 Italian universities with at least 29,000 students enrolled. Medium-sized universities are defined as those with between 10,000 and 29,000 enrolled students, while small universities have fewer than 10,000 enrolled students.

[6] Doctoral theses defended in Italy must be deposited at the National Central Libraries of Rome and Florence by law.

more general limitation is that the analysis does not include the funding to HEIs from two agencies responsible for research on climate topics, namely the *National Agency for New Technologies, Energy and Sustainable Economic Development* (ENEA) and the *Italian Institute for Environmental Protection and Research* (ISPRA). The possibility of identifying the role of academic research in the funding instruments managed by ENEA and ISPRA is currently limited.

Italy's Political Approach to Climate Change

Despite its vulnerability to the effects of climate change, Italy has long lacked international political leadership in addressing the issue and devotes less attention to it than other Western countries (Westerhoff, 2010). Starting with the Kyoto Protocol,[7] the country has maintained a political approach based primarily on joining international agreements and has set its climate policy in accordance with EU strategies. In recent years, national policies have undoubtedly become more concerned with climate change in response to increased international attention on the issue, but there is still no national climate law in place, as in the United Kingdom, Germany, and France,[8] which would provide a unified regulatory framework based on scientific evidence (Duwe & Evans, 2020).

In Italy, climate-related regulations are under the responsibility of the government, parliament, and related bodies,[9] but regional and local governments play an increasingly important role (OECD, 2013). In this regard, experts argue that the Italian multilevel governance system on climate is not well-coordinated among institutions (Pietrapertosa et al.,

[7] The Kyoto Protocol, ratified by Italy in 2002, was the first legally binding operational instrument establishing specific targets for reducing greenhouse gas emissions. Italy pledged to reduce greenhouse gas emissions by 6.5% compared to 1990 levels between 2008 and 2012.

[8] In the UK, the Climate Change Act, adopted in 2008; in France, the Energy Transition Green Growth Act (*Loi de transition énergétique pour la croissance verte*), adopted in 2015; in Germany, the Climate Protection Law (*Klimaschutzgesetz*) adopted in 2019. Other European countries have adopted national climate laws, including Denmark, Finland, Ireland, the Netherlands, and Sweden. Among Mediterranean countries, Spain has recently adopted a national climate law in 2021.

[9] The Ministries of Environment and Infrastructure have the primary responsibility of promoting ecological transition in the country, while inter-ministerial committees are tasked with coordinating actions and approving national plans and strategies on matters that affect multiple ministries.

2021; Colafrancesco & Di Mambro, 2023). The *Climate Change Performance Index*, developed by Germanwatch, NewClimate Institute, and Climate Action Network, which analyses and compares climate change mitigation policy efforts across 63 countries and the EU, shows, in the 2024 edition, Italy performing poorly, ranking 44th and having worsened significantly compared to the previous year (Burck et al., 2023).[10] The news of the publication of the report was remarked on by environmental associations and received some media coverage but garnered little attention from the political parties currently in office.

Evolution of National Policy on Climate and Energy

After the ratification of the Kyoto Protocol in 2002, Italy's first notable step toward acknowledging the significance of climate change on the political agenda was the organization of a *National Conference on Climate Change* (Conferenza Nazionale sui Cambiamenti Climatici) by the Ministry of the Environment in 2007. Following the event, a *Climate Manifesto* (Manifesto per il clima) was released, framing climate change as a national issue and emphasizing the importance of coordinating efforts to both mitigate and adapt to its effects.[11] The first political initiatives to address the new priority focused on the energy sector, with the goals of achieving Kyoto Protocol targets, promoting renewable energy, and improving energy efficiency. In this regard, in 2008, the *National Energy Strategy* (Strategia Energetica Nazionale, SEN) was introduced into the legal system and launched five years later, mapping a long-term decarbonization path for the country (IEA, 2016). In addition, the *National Action Plan for Greenhouse Gas Reduction* was revised in 2013, almost ten years after it was drafted, to include the EU Member States' commitments under the *Effort Sharing Decision* (Pantaleoni, 2013).

In 2015, with the approval of the *National Climate Change Adaptation Strategy* (Strategia Nazionale di Adattamento ai Cambiamenti Climatici, SNAC), Italy solidified its commitment to a climate policy aimed at creating a national vision based on the vulnerabilities of natural systems and

[10] In the 2024 edition of the Index, the authors decided that none of the countries could be ranked in the top three, since none of them "is doing enough to prevent dangerous climate changes."

[11] https://www.isprambiente.gov.it/it/archivio/eventi/anno-2007/conferenza-cccn2007 (accessed June 30, 2023).

considering all pertinent socio-economic sectors. The document was drafted in response to the *EU Strategy on Climate Change Adaptation* (EC, 2013), which encouraged EU Member States to integrate adaptation measures into various national and local policies based on vulnerability profiles and resilience capacity. The SNAC made several recommendations, including that climate impacts be considered from the beginning of the development process of each energy project and that climate data collection be improved to enhance the accuracy of projections regarding the potential impacts of climate change (MATTM, 2015). To enable the implementation of the SNAC, the Italian government began drafting a *National Climate Change Adaptation Plan* (Piano Nazionale di Adattamento ai Cambiamenti Climatici, PNACC) in 2016, which analyzed expected climate change impacts, identified specific vulnerabilities, and outlined a series of adaptation measures for the energy sector. However, the plan was finally approved only in 2023,[12] becoming operational as of February 2024. At the end of 2017, the Ministry of Environment approved another strategic document, the *National Sustainable Development Strategy* (Strategia Nazionale per lo Sviluppo Sostenibile, SNSvS), which set out strategic choices and objectives aligned with the 2030 Agenda's Sustainable Development Goals, including the 13th goal of the agenda, which urges taking immediate measures to mitigate the effects of climate change.[13]

Since 2020, the Italian regulatory framework on climate-related issues has become more complex and stratified. That year, the first *Italian National Integrated Energy and Climate Plan* (Piano Nazionale Integrato per l'Energia e il Clima, PNIEC) was released.[14] The plan outlined the country's vision for its energy future up to 2030 with a focus on environmental, social, and economic sustainability, comprising five integrated lines of action, including research and innovation. Then, in 2021, the *Italian Long-Term Strategy on the reduction of greenhouse gases* was published, setting decarbonization objectives for 2050.[15] Furthermore, the *National Recovery and Resilience Plan* (NRRP), approved in 2021 in response to the COVID-19 crisis, included measures to improve climate

[12] https://www.mase.gov.it/notizie/clima-approvato-il-piano-nazionale-di-adattamento-ai-cambiamenti-climatici (accessed January 8, 2024).

[13] https://www.mase.gov.it/pagina/la-snsvs (accessed June 30, 2023).

[14] https://www.mase.gov.it/comunicati/pubblicato-il-testo-definitivo-del-piano-energia-e-clima-pniec (accessed June 30, 2023).

[15] https://www.mase.gov.it/sites/default/files/lts_gennaio_2021.pdf (accessed June 30, 2023).

resilience.[16] Lastly, the *Plan for Ecological Transition* (Piano per la Transizione Ecologica, PTE) was created as a coordinating and planning tool for the various measures outlined in the *European Green Deal*, a package of strategic initiatives to put the EU on the path to a green transition (EC, 2019), and in the NRRP regarding the ecological transition.

However, despite increased efforts in recent years to address the climate crisis, some observers have expressed overall concern about a lack of internal alignment between the various Italian climate-related policies and have proposed the creation of a unified national regulatory instrument, the coordination of governance, and enhanced scientific support to guide actions based on the most recent evidence on climate change (Colafrancesco & Di Mambro, 2023). In October 2023, the environmental associations WWF, Greenpeace, Legambiente, Kyoto Club, and Transport and Environment promoted a framework climate law for Italy. Most of the support for this law came from the opposition parties in the Italian parliament, but it was also presented by representatives of the majority government parties.[17] The law would provide for, among other things, the establishment of a Scientific Climate Council, a Citizens' Assembly that could propose programs and strategies for climate policies, and a mechanism providing for the calculation of the updated total carbon budget for each economic sector on a triennial basis.

Italian R&D Policies Addressing Climate Change

Almost all Italian climate and energy policy documents stress the importance of investing in R&D and innovation, and some strategies and plans, such as the SNAC and SEN, have been developed in collaboration with the scientific community. The SNAC has drawn attention to the value of collaboration between policymakers, stakeholders, and scientists in addressing potential future scenarios of climate change, emphasizing the importance of steering scientific research toward producing climate risk

[16] Particularly, the "Green Revolution and Green Transition" mission, which constitutes the largest share of the NRRP resources, accounting for over 31% of the total amount (about 70 billion euros), is dedicated to intensifying Italy's commitment to achieving energy efficiency, building upgrades, sustainable mobility, infrastructure enhancement, renewable energy utilization, promotion of industrial and hydrogen supply chains, and digitization of the grid infrastructure (Camera dei Deputati, 2021).

[17] https://www.wwf.it/pandanews/ambiente/italia-via-a-legge-su-clima/ (accessed October 16, 2023).

analyses tailored to vulnerable sectors such as infrastructure, agriculture, urban settlements, transportation, businesses, and energy (MATTM, 2015). The SEN, on the other hand, sought to analyze the R&D situation in its field, noting weak R&D in the energy sector and setting a goal of doubling R&D investments in clean energy (MISE and MATTM, 2017).

The importance of funding R&D projects at universities and public research bodies, as well as collaborations between the public and private sectors, has been repeated in the *National Research Program* (Programma nazionale per la ricerca, PNR), the periodic strategy document for public research written by the MUR,[18] which reports the government's guidelines for scientific and technological policy. However, recent analyses tell us that Italy is a country that devotes a very small portion of its government R&D budget allocations for R&D (GBARD)—approximately 10%—to project funding, promoting a limited number of funding instruments (Reale, 2018), which place less strategic emphasis than other European countries on societal and technological goals in terms of total investments and diversification over the last decade (Spinello et al., 2023).

Climate-Related R&D Project Funding for Academic Research and the Public Sector

The analysis of data from the EFIL dataset painted a comprehensive picture of portfolio of the Italian public project funding instruments, which we used to identify schemes potentially aimed at climate change research. The research instrument that the SNAC has labeled the "main financial support for climate research" (MATTM, 2015) is the *Special Supplemental Fund for Research* (Fondo Integrativo Speciale per la Ricerca, FISR), which is a broad-topic scheme that funds very few select research initiatives of strategic importance and high complexity, according to the PNR. In terms of climate change-related research, the FISR contributed to the financing of one specific project in 2001—CLIMESCO, on the evolution of cropping systems due to climate change—and to the establishment of three research networks on climate change and energy research in 2005, 2015, and 2017. The first was the *Euro-Mediterranean Center on Climate Change* (Centro Euro-Mediterraneo sui Cambiamenti Climatici,

[18] MUR takes the lead in managing and maintaining oversight of project funding instruments for Italian HEIs and the public sector in general, calling for proposals for basic and targeted research.

CMCC),[19] a consortium of entities, including universities and research centers, established with the aim of generating simulations, diagnoses, and forecasts on climate change. The second was the *International Centre of Advanced Study in Environment, Ecosystem and Human Health* (Centro Internazionale di Studi Avanzati su Ambiente, ecosistema e Salute umana, CISAS), whose main proponent was the National Research Council of Italy. The third regarded the funding for the creation of the *National Technology Cluster on Energy* (Cluster Tecnologico Nazionale Energia), a public-private partnership led by a public sector agency—ENEA—to create opportunities for technological and innovative development related to energy in the industrial sector through combined research efforts. Although the FISR funding scheme could include universities among the beneficiaries, over time, public research institutions have made up the majority of its beneficiaries. Taking the years between 2014 and 2017 as an example, it financed only 28 projects, only seven of which were proposed by universities (and only one related to climate change research), while public research agencies accounted for the lion's share. Given these characteristics, it is difficult to gain a complete picture of universities' mobilization capacity with respect to the opportunities provided by this program, which is for the public sector in general rather than academia specifically.

One Italian funding instrument for the public sector, designed regardless of the emergence of climate change studies as a current trend but focusing on topics potentially related to it, is the long-standing *National Antarctic Research Program* (Programma Nazionale di Ricerca in Antartide—PNRA). Created in 1985, PNRA is a historic funding program that provides for the organization of scientific expeditions to Antarctica, with activities to be carried out at scientific stations to conduct research in the Antarctic region, focusing on topics such as biodiversity and adaptation, biology, and global change. Surprisingly, this initiative is almost never mentioned in climate-related policy strategies, perhaps due to the global reach of the studies, which may or may not have an impact on national policies. Again, in this case, the characteristics of the program in terms of the different types of eligible beneficiaries prevent us from clearly identifying the role that universities play in project leadership. Furthermore, the underlying theme of the program makes it difficult to identify projects more markedly linked to climate change.

[19] Website: https://www.cmcc.it/

In Italy, government-funded academic research on climate change has largely been conducted within the broad-themed *PRIN program*, the research policy instrument that allocates most of the project funding for the public sector in Italy (Spinello et al., 2023) and whose participants are overwhelmingly universities (98% from 1999 to 2015 according to NATPRO data). Indeed, the *PRIN program* is the main competitive funding instrument for basic academic research and supports two- or three-year collaborative projects (depending on the year of the calls) in all fields of science. Academics are free to determine the content, themes, and methods, competing with proposals across all disciplines. In this regard, the program contributes to opportunities for HEIs to undertake spontaneous initiatives on specific issues to address in their projects, including climate change. The guiding principles of the program are that the project coordinator and operational unit managers must have a significant background in science, and the funded project must be original and have a scientific impact. The successive calls for proposals, which were annual until 2010 and then every two years after that (or three years in 2015), have highlighted changes in the evaluation criteria of the proposals, especially starting from the 2017 call, and in the selection of the proposals, the bibliometric criterion was introduced alongside the traditional peer review.

In short, the overall stimulus for climate change research in terms of project funding opportunities for universities is limited to a very small number of instruments and primarily focuses on competition for resources within a generalist funding scheme for a "wide range of research" such as the *PRIN program*. Research on climate change is therefore mostly left to the initiative of the universities that apply for funds from that program.

Climate Change Projects Funded Under the PRIN Program

By combining EFIL data with NATPRO data, we have carried out an analysis of the projects on climate change conducted by universities funded under the *PRIN program* over six consecutive calls launched between 2009 and 2020. The PRIN calls have had a particularly irregular trend over time, and the amount of funding has varied greatly. Consider that the 2009 and 2010–2011 earmarks came to approximately €105 and €170 million, respectively, while the 2012 allocation was only €32.3 million and the 2015 allocation was €91.9 million. This amount increased to €391 million in the 2017 call, and the 2020 call provided for a total of approximately €179 million.

Using the list of keywords selected for the analysis (see par. 4.3), we found that the *PRIN program* funded a total of 92 out of 2207 (4.2%) projects on climate change topics in the selected period, never exceeding 20 projects per call and ranging from a minimum of nine in the 2012 call to a maximum of 20 in the 2017 call.

As can be seen in Fig. 4.1, the number of total projects approved varies widely over the calls, and this also influences the number of climate change-themed projects. Five or six projects per round present the keywords most strongly linked to climate change, except for the 2010/2011 call, which funded 10. Curiously, it is precisely in the years in which fewer projects are funded that the percentage of the budget allocated shows the higher values (in the 2010/2011 call, 8.3% of the funded budget, about €2.5 million; in the 2012 call, 7.8%, about €6 million). The last calls present a decline in percentage terms in the budget reserved for climate change

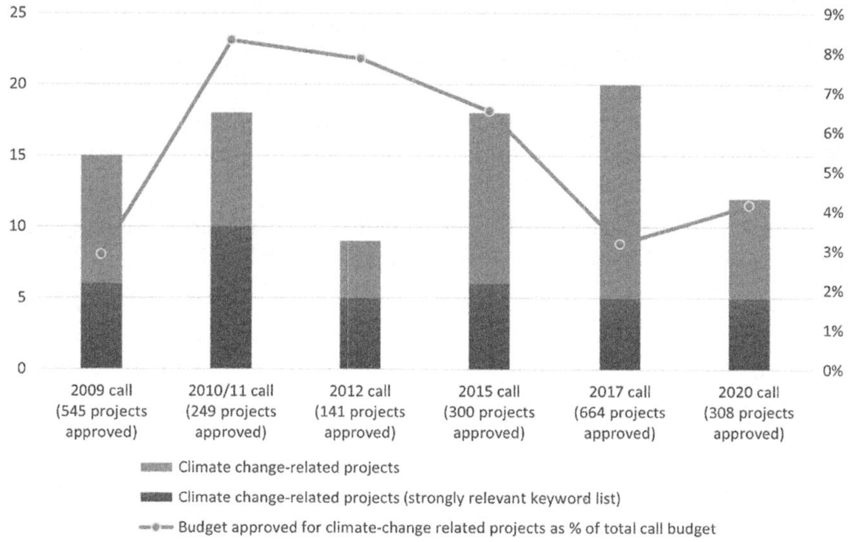

Fig. 4.1 Number of climate change-related PRIN projects by call and percentage of the total call budget obtained. (Source: authors' processing of RISIS EFIL and NATPRO data combined and integrated with data from the PRIN calls for proposals portal of the research general directorate of the MUR)

topics (in the 2015 call, 6.5%, €6 million; in the 2017 call, 3.2%, €12.4 million), with slight growth in the last call (4.1%, €7.4 million). Overall, the data show that this topic is of interest to applicants, and while the total number of projects is not large, the budget reserved, given the multidisciplinary and open-topic nature of the *PRIN program*, is considerable.

Climate change-themed projects fall into all three ERC domains, demonstrating the multidisciplinary nature of the topic and the interest that HEIs may devote to specific aspects across multiple subjects. The Physical, Earth, and Engineering Sciences domain shows clear dominance; within this macro-sector, ample space is given to projects in PE8—Products and Processes Engineering (19 projects) and PE10—Earth System Science (15 projects). In the Life Science domain, Environmental Biology and Applied Life Science account for a total of 23 projects. It is interesting to note that the Social Sciences and Humanities domain makes up 12% of all projects, which are mostly related to the sub-domain Environment, Space, and Population.

Table 4.1 shows which HEIs are most active in PRIN-funded climate change research. Three universities stand out above the others: Florence and Padua (with 22 participants each) and Bologna (20). Except for the University of Perugia, the top ten universities in this ranking are all large.

Figure 4.2 expands the analysis of the HEIs' participation to the regional level. The interest in climate change topics spans the peninsula from North to South. Despite the predominance of participation in

Table 4.1 Top ten Italian HEIs by the number of climate change-related PRIN projects (total participation from the 2009 call to the 2020 call)

HEI	*Region*	*Size*	*Participations*
University of Florence	Tuscany	Large	22
University of Padua	Veneto	Large	22
University of Bologna	Emilia-Romagna	Large	20
Sapienza University of Rome	Lazio	Large	19
Polytechnic University of Milan	Lombardy	Large	18
University of Palermo	Sicily	Large	18
University of Turin	Piedmont	Large	16
University of Pisa	Tuscany	Large	14
University of Perugia	Umbria	Medium	13
Polytechnic University of Turin	Piedmont	Large	12

Source: Authors' processing of RISIS EFIL and NATPRO data combined and integrated with data from the PRIN calls for proposals' portal of the research general directorate of the MUR

Lombardy, with the Polytechnic University of Milan taking the lion's share, and Lazio, with Sapienza University of Rome accounting for many projects, the regions in the South and the Islands are very active, with a total of 138 participations. Particularly, the universities in Sicily and Campania account for 38 and 36, respectively.

As displayed in Fig. 4.3, large HEIs clearly drive participation over the entire period considered, accounting for between one half and two thirds of total projects. This is not surprising given the established intensity of their research and their reputation, which allows them greater access to external funding sources, making them more competitive in securing funding for larger-scale projects and initiatives. However, the interest of medium-sized HEIs (41% in the 2015 call and 33% in the 2020 call) shows percentage growth over time.

Ultimately, it is possible to observe that with limited and variable funding, and thus opportunities that are not specifically ***intended*** for climate

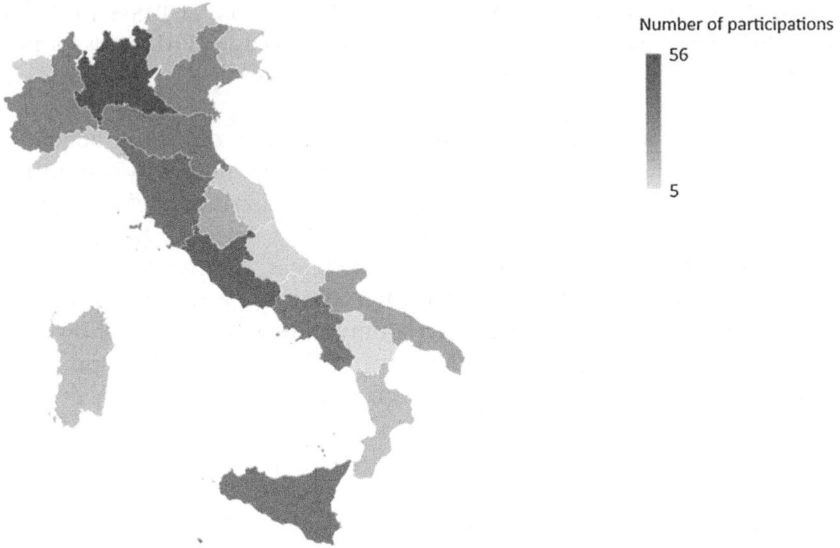

Fig. 4.2 Distribution by region of Italian HEIs participating in climate change-related PRIN projects (total participation from the 2009 call to the 2020 call). (Source: authors' processing of RISIS EFIL and NATPRO data combined and integrated with data from the PRIN calls for proposals portal of the research general directorate of the MUR)

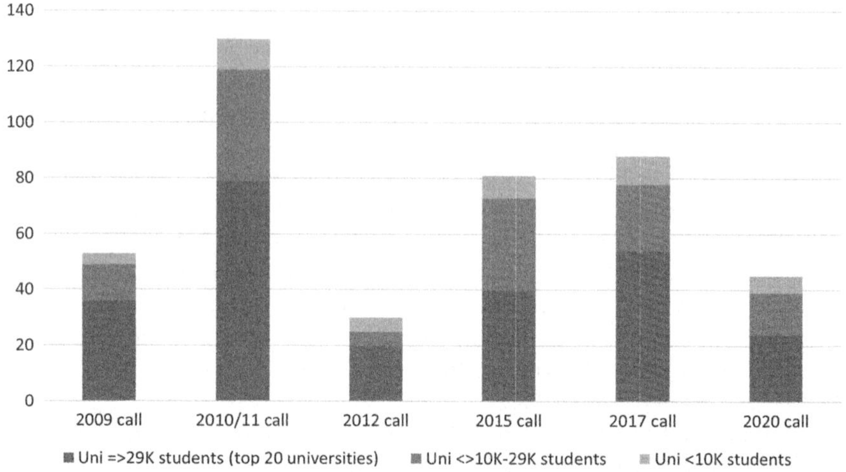

Fig. 4.3 Participations in climate change-related PRIN projects by university size for the 2009–2020 calls. (Source: authors' processing of RISIS EFIL and NATPRO data combined and integrated with data from the PRIN calls for proposals portal of the research general directorate of the MUR)

change studies and are *provided* within a generalist offer, the beneficiaries—Italian HEIs—somewhat *mobilize* these opportunities by submitting applications for the funding of projects. The mobilization of efforts results in a fair amount of success in terms of the budget received. Research proposals are submitted in a variety of thematic areas, but the majority are in Engineering and Earth Sciences. The interesting fact is that HEIs respond from North to South, with southern universities of all sizes showing a strong interest in work on the topic, as perhaps they are the most interested in the implications for the Mediterranean region.

Doctoral Theses on Climate Change

Finally, we have analyzed data on PhD theses defended at Italian universities to gather further evidence of the activism of academic research on the issue of climate change, regardless of research policy incentives. According to statistics from the MUR from 2010 to 2019, about 113,000 doctoral theses were defended, and our analysis identified 1036 theses related to climate change. Figure 4.4 shows the trend in doctoral theses on this

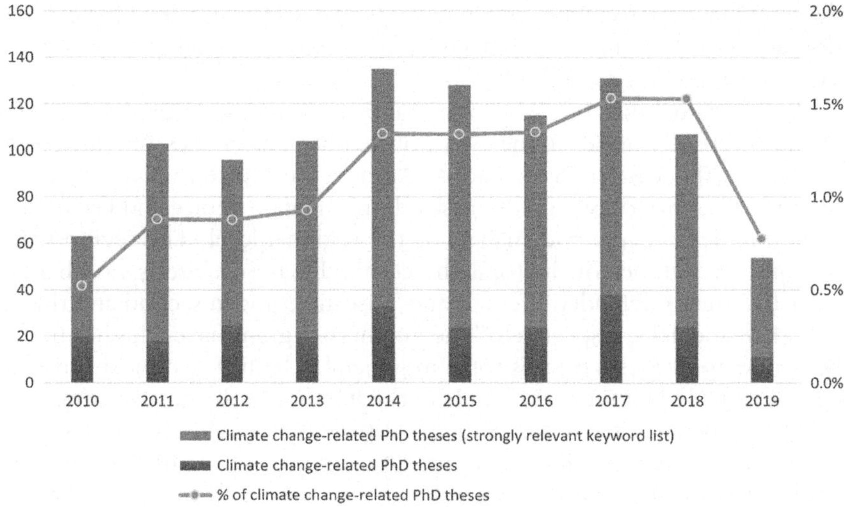

Fig. 4.4 Number of climate change-related PhD theses defended at Italian universities, and their percentage on the total of PhD theses defended in Italy by year from 2010 to 2019. (Source: authors' processing of BNCR data on PhD theses defended in Italy)

topic. The data suggest a general upward trend in the number of climate change-related theses until 2017, with some annual fluctuations. The number of theses steadily increased from 63 in 2010 to a high of 135 in 2014. There was a significant decrease in the number of theses after 2017, with only 54 identified in 2019. Several factors, including student interest, research trends, available funding, and academic priorities, may influence annual fluctuations. If we limit the analysis to the strongly related keyword list, we observe some variations over time, with no clear upward or downward trend. It is worth noting, however, that the number of theses identified using the short list of keywords increased from 20 in 2010 to 33 in 2014. Over the years considered, the percentage of climate change-related theses in relation to the total number of theses in the database increased steadily to 1.5% in 2017 and 2018, before falling to 0.8% in 2019.

About 71% of PhD theses fall into the ERC domain of Physical Sciences and Engineering, with Life Science and Social Science and Humanities accounting for 18% and 10% of the total, respectively. The Physical Sciences and Engineering domain shows a steady increase over the years from 50% in 2010 to 88% in 2019. As for the Social Science and Humanities domain, the data show a relatively stable variation, ranging from 6% to 16% over the years considered.

As shown in Table 4.2, most doctoral theses on topics related to climate change were discussed at the two polytechnic universities of Turin and Milan and the University of Padua. Among the top ten universities, we note the presence of two medium-sized universities (Venice and Calabria).

Figure 4.5 expands the analysis to the regional level. The region with the highest number of doctoral theses on climate change is Lombardy, with 142 theses defended. Veneto and Tuscany come in second and third, with 128 and 103, respectively. The northern regions have a higher number of doctoral theses on this topic in general. The first and most obvious reason is the large number of HEIs in the north, but the greater complexity of the thematic portfolio of doctoral programs might also play a role. Other factors that may influence the data and require further exploration could be student interest in addressing these issues and the availability of funding (e.g., according to the EU Open Data Portal, the universities of

Table 4.2 Top ten Italian HEIs by the number of climate change-related PhD theses defended (2010–2019)

University	*Region*	*Size*	*PhD theses on topics related to climate change*
University of Padua	Veneto	Large	71
Polytechnic University of Turin	Piedmont	Large	66
Polytechnic University of Milan	Lombardy	Large	60
University of Bologna	Emilia-Romagna	Large	55
University of Florence	Tuscany	Large	55
Ca' Foscari University of Venice	Veneto	Medium	47
University of Genoa	Liguria	Large	42
University of Naples Federico II	Campania	Large	40
University of Milan	Lombardy	Large	37
University of Calabria	Calabria	Medium	36

Source: Authors' processing of BNCR data on PhD theses defended in Italy

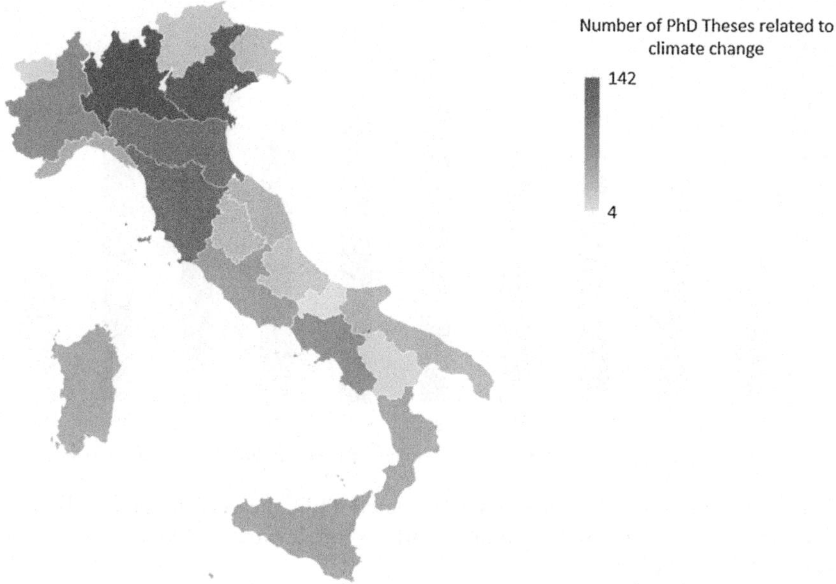

Fig. 4.5 Distribution by region of Italian HEIs in which PhD theses related to climate change have been defended (2010–2019). (Source: authors' processing of BNCR data on PhD theses defended in Italy)

Bologna, Venice, and Padua were involved in 45% of the projects on climate change with Italian participants funded under EU-H2020[20]).

Figure 4.6 illustrates the distribution of doctoral theses on climate change by university size. Larger HEIs play a dominant role but show a steady decline in subsequent years. This trend could reflect a change in student interest or research priorities within the larger HEIs. In the last year considered, medium-sized HEIs outnumber large ones in terms of the number of climate change-related theses defended.

Overall, the number of doctoral theses on climate change has not been high. This suggests that the topic of climate change may not be a high priority in doctoral programs in Italy. However, an upward trend, driven mainly by HEIs in northern Italy, was observed at least until 2017,

[20] Data source: https://emiliaromagnainnodata.art-er.it/partecipazione-delle-universita-italiane-ad-horizon-2020/ (accessed June 8, 2023).

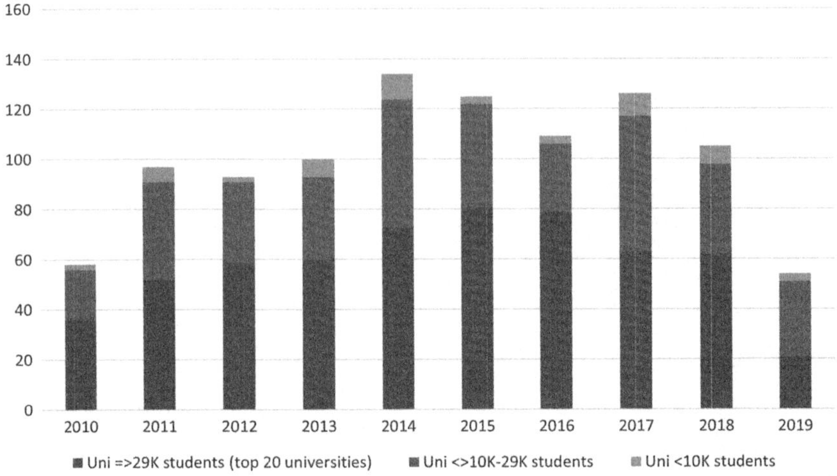

Fig. 4.6 Number of climate change-related PhD theses defended in Italy by university size (2010–2019). (Source: authors' processing of BNCR data on PhD theses defended in Italy)

indicating a growing interest in correspondence with increased European attention (and funding opportunities) on these issues. While the percentage reduction after 2017 is more difficult to explain, the drop in absolute terms may be related to a drop in the number of doctorates earned in Italy over the next two years, with 18% fewer doctorates awarded in 2019 compared to 2017.[21]

Discussion and Conclusions

Policymakers at national and supranational levels are increasingly committed to encouraging action on climate change research by setting objectives and providing funding to encourage the development of sustainable solutions (see Chap. 1 of this book). As key actors in the generation of knowledge, HEIs are called upon to offer expertise to facilitate the green and digital transition toward a future that operates within the limits of the planet's natural resources. This request can produce deep transformations

[21] Source: MUR—Higher Education Data Portal (Portale dei dati dell'istruzione superiore) at https://dati-ustat.mur.gov.it/ (accessed June 8, 2023).

in the HEIs, and these transformations can also affect their attitudes about the directionality of science in solving social problems (Kaldewey, 2018; Kuhlmann & Rip, 2018). Thus, research policies may effectively steer the actors toward addressing these issues by prioritizing them within R&D strategies and by creating specific instruments (Witte, 2023). However, the mere existence of specific instruments does not ensure that potential beneficiaries will act upon them, especially in the presence of fuzzy political guidance. It is therefore necessary to consider not only the availability of funding mechanisms but also how scholars or research groups perceive the potential incentives stemming from research policies (Reale et al., 2014). The latter aspect, relating to the effectiveness of mechanisms driving the directionality of research activities, has been fairly underestimated in the literature.

In this study, the authors applied a methodological approach based on the framework of both proposed and perceived R&D funding opportunities to the specific case of Italy, a country that is significantly vulnerable to the risks of climate change. They examined the broader context of the Italian political approach on the topic of climate change and focused on the R&D funding opportunities accessible to Italian HEIs that have been *intended and provided* for climate change research in recent years. Further, they analyzed how universities have *perceived* these opportunities and how they have consequently *reacted*.

The evidence revealed that—despite a multitude of policy documents (strategies and plans on climate and energy, such as the SNAC and SEN) and the ministerial *National Research Program* (PNR) highlighting the importance of science in tackling climate change—Italian research policies have offered very limited opportunities for funding *intended* for climate change research. The FISR scheme, which the SNAC has labeled the "main financial support for climate research", took the form of an opportunity for a small number of beneficiaries and was not specifically (or exclusively) for academic research and projects on climate change. Even the PNRA, the *National Antarctic Research Program*, is difficult to classify, as although it may also fund research on climate change, it is strongly oriented toward specific studies on the Antarctic region and accessed by research organizations from the public sector in general.

In the absence of funding schemes specifically intended for climate change and targeting HEIs, universities must consider opportunities that are provided as part of a generalist offer of funding for academic research. Thus, academics from Italian HEIs have attempted to capitalize on the limited possibilities for obtaining funding for projects on climate change issues by submitting proposals to the largest Italian project funding scheme—the *PRIN program*. The mobilization of efforts has yielded a reasonable amount of success in terms of the budgets awarded. The funded climate change-related research projects have covered a wide range of thematic areas, with Engineering and Earth Sciences receiving the most attention. It is interesting that HEIs have responded from North to South, with a particular role played by larger universities, and the universities of Florence and Padua showing a strong interest in the study of climate change-related topics.

The study also investigated the propensity of HEIs to self-select climate change research as a topic of interest, regardless of the government incentives, analyzing the doctoral theses defended at Italian universities. The findings showed the sporadic activism of PhD students toward climate change issues in the context of doctoral programs. Northern Italian universities appear to be the best equipped to address these issues with their doctoral students, while universities in southern Italy appear to lag behind. Furthermore, medium-sized universities appear to be more inclined in recent years to assign doctoral theses on climate change than larger universities.

In summary, the answers to our research questions are as follows:

Considering Italy's position as a vulnerable country and the interest of universities, the study's findings emphasize the lack of dedicated funding schemes addressing climate change in the national context, which is a major obstacle for universities to engage in comprehensive studies on this pressing issue. However, this circumstance does not deter Italian HEIs from conducting studies on the topic, taking advantage of a generalist funding program and expressing a certain amount of their research potential.

The funding has primarily supported short-term projects (typically lasting 24–36 months) which fall short of effectively addressing the complex challenges that society faces. To address this gap, a more structured and comprehensive approach is required, with long-term financial resource mobilization supplemented by follow-on funding.

New calls for competitive R&D programs launched in 2022 with NRRP funds show an overall reconfiguration of the project-based funding instruments in the Italian context, in the name of the introduction of mechanisms and priorities aimed at harmonizing the research system with European Union initiatives on research and innovation. For example, in the "Extended Partnerships" call, the funded networks and their research activities must be focused on specific topics, including environmental issues. Furthermore, the "National Centers" call is financing one national center on the theme of biodiversity, with universities playing a role.

These new initiatives under the NRRP are obviously highly promising. However, it is crucial to bear in mind that climate change is a multidisciplinary and highly complex field of research that is interwoven and overlapping with many other research areas, so it will be necessary to guarantee opportunities across many different disciplines. Adequate funding for multidisciplinary climate change research can improve our understanding of the causes of phenomena that occur in Italy and pave the way for novel technologies, practices, and regulations to mitigate risks and tackle the major issues.

Future studies may analyze the characteristics, outcomes, and long-term sustainability of the initiatives facilitated by the NRRP to understand whether there have been transformative changes in the government's steering capability toward climate change solutions. Furthermore, one interesting route for further research is to investigate how government policies can drive other actors and stakeholders, particularly shedding light on the potential emergent role of nonprofit organizations.

Acknowledgments This work was supported by RISIS, EU grant number 824091, and the project "Fostering Open Science in Social Science Research" (FOSSR), funded by the European Union—NextGenerationEU under NRRP Grant agreement n. MUR IR0000008. The content of this chapter reflects only the authors' view. The European Commission and MUR are not responsible for any use that may be made of the information it contains.

References

Barbante, C., Carraro, C., Navarra, A., Pasini, A., & Valentini, R. (2022). *Lettera aperta degli scienziati del clima alla politica italiana (Open letter from climate scientists to Italian policymakers)*. Retrieved June 30, 2023, from https://www.greenandblue.it/2022/08/03/news/crisi_climatica_lettera_aperta_scienziati_politica_italiana-360156752/

Burck, J., Uhlich, T., Hohne, N., Nascimento, L., & Bals, C. (2023). *Climate change performance index 2024*. Germanwatch, NewClimate Institute & Climate Action Network.

Camera dei Deputati. (2021). *Cambiamenti climatici - Ambiente e gestione del territorio - Politica estera e questioni globali. Documentazione parlamentare*

(Climate change - Environment and land management - Foreign policy and global issues. Parliamentary Documents). Retrieved June 30, 2023, from https://temi.camera.it/leg18/temi/tl18_cambiamenti_climatici.html

Castellari, S., Venturini, S., Ballarin Denti A., Bigano A., Bindi M., Bosello F.... Zavatarelli M. (Eds.) (2014). *Rapporto sullo stato delle conoscenze scientifiche su impatti, vulnerabilità ed adattamento ai cambiamenti climatici in Italia (Report on the state of scientific knowledge on impacts, vulnerability and adaptation to climate change in Italy)*. Ministero dell'Ambiente e della Tutela del Territorio e del Mare, Roma.

Colafrancesco, G., & Di Mambro, C. (2023). *Una governance per il clima in Italia. Quali elementi per una legge quadro per il clima (Climate governance in Italy. What elements for a climate framework law)*. Technical Report Gennaio 2023. ECCO The Italian Climate Change Think-Tank.

Duwe, M., & Evans, N. (2020). *Climate laws in Europe: Good practices in Net-Zero management*. Ecologic Institute.

EC. (2013). *An EU strategy on adaptation to climate change. Communication from The Commission to The European Parliament, The Council, The European Economic and Social Committee and The Committee of the Regions*. COM(2013) 216 final, Brussels.

EC. (2019). *The European Green Deal. Communication from The Commission to The European Parliament, The European Council, The Council, The European Economic And Social Committee and The Committee of the Regions*. COM(2019) 640 final, Brussels.

Foray, D., Mowery, D. C., & Nelson, R. R. (2012). Public R&D and social challenges: What lessons from mission R&D programs? *Research Policy, 41*(10), 1697–1702.

Guthrie, S., Wamae, W., Diepeveen, S., Wooding, S., & Grant, J. (2013). *Measuring research. A guide to research evaluation frameworks and tools*. RAND Europe.

IEA. (2016). *Energy policies of IEA countries: Italy 2016 review*. Accessed June 30, 2023, from https://www.iea.org/reports/energy-policies-of-iea-countries-italy-2016-review

IPCC. (2022). *Climate change 2022: Impacts, adaptation and vulnerability. Contribution of Working Group II to the Sixth Assessment Report of the Intergovernmental Panel on Climate Change*. Cambridge University Press, Cambridge, UK and New York, NY, USA.

Kaldewey, D. (2018). The grand challenges discourse: Transforming identity work in science and science policy. *Minerva, 56*(2), 161–182.

Kuhlmann, S., & Rip, A. (2018). Next generation innovation policy and grand challenges. *Science and Public Policy, 45*(4), 448–454.

Lepori, B., & Reale, E. (2012). S&T indicators as a tool for formative evaluation of research programs. *Evaluation, 18*(4), 451–465.

Lepori, B., & Reale, E. (2019). The changing governance of research systems. Agencification and organizational differentiation in research funding organizations. In D. Simon, S. Kuhlmann, J. Stamm, & W. Canzler (Eds.), *Handbook on science and public policy* (pp. 448–463). Edward Elgar Publishing.

Lepori, B., van den Besselaar, P., Dinges, M., Potì, B., Reale, E., Slipersæter, S., Thèves, J., & van der Meulen, B. (2007). Comparing the evolution of national research policies: What patterns of change? *Science and Public Policy, 34*(6), 372–388.

Lionello, P., & Scarascia, L. (2018). The relation between climate change in the Mediterranean region and global warming. *Regional Environmental Change, 18*(5), 1481–1493.

Macchiati, A. (2009) *Le politiche contro il cambiamento climatico nell'Unione Europea e in Italia. Nota elaborata per il gruppo di studio ASTRID "Clima, energia e ambiente" (Climate change policies in the European Union and Italy. Note prepared for the ASTRID Study Group on Climate, Energy and the Environment).* Retrieved June 30, 2023, from https://www.astrid-online.it/static/upload/protected/Macc/Macchiati_cambiamento-climatico.pdf

MATTM. (2015). *Strategia Nazionale di Adattamento ai Cambiamenti Climatici (National Climate Change Adaptation Strategy).* Ministero dell'Ambiente e della Tutela del Territorio e del Mare. Retrieved June 30, 2023, from https://www.mase.gov.it/sites/default/files/archivio/allegati/clima/strategia_adattamentoCC.pdf

Mazzucato, M. (2018). Mission-oriented innovation policies: challenges and opportunities. *Industrial and Corporate Change, 27*(5), 803–815.

MISE and MATTM. (2017). *Strategia Energetica Nazionale (SEN) 2017 (National Energy Strategy (SEN) 2017).* Ministero dello Sviluppo Economico e Ministero dell'Ambiente e della Tutela del Territorio e del Mare. Retrieved June 30, 2023, from https://www.mimit.gov.it/images/stories/documenti/Testo-integrale-SEN-2017.pdf

OECD. (2013). *OECD environmental performance reviews: Italy 2013.* OECD Publishing.

Overland, I., & Sovacool, B. K. (2020). The misallocation of climate research funding. *Energy Research and Social Science, 62*(2020), 101349.

Pantaleoni, M. (2013). *Italy climate policy progress report.* Submitted to the European Commission pursuant to Decision No 280/2004/EC, Article 3(2). Retrieved June 30, 2023, from https://groupware.sinanet.isprambiente.it/reportnet/library/projected_emissions/delivery-2013/it_2013-italy-climate-policy-progress-report

Pedersen, D. B., Gronvad, J. F., & Hvidtfeldt, R. (2020). Methods for Mapping the impact of social sciences and humanities—A literature review. *Research Evaluation, 29*(1), 4–21.

Pietrapertosa, F., Salvia, M., De Gregorio Hurtado, S., Geneletti, D., D'Alonzo, V., & Reckien, D. (2021). Multi-level climate change planning: An analysis of the Italian case. *Journal of Environmental Management, 289*(2021), 112469.

Reale, E. (2018). Il finanziamento pubblico per Ricerca e Sviluppo. Le politiche dell'Italia in prospettiva comparata (R&D public funding in Italy. Italy's policies in comparative perspective). In Consiglio Nazionale delle Ricerche (ed.), *Relazione sulla ricerca e l'innovazione in Italia. Analisi e dati di politica della scienza e della tecnologia. Prima edizione (Report on Research and Innovation in Italy. Analysis and data on Science and technology policy* (1st edn, pp. 225–253). CNR Edizioni.

Reale, E., Nedeva, M., Thomas, D., & Primeri, E. (2014). Evaluation through impact: a different viewpoint, *Fteval Journal for Research and Technology Policy. Evaluation, 39*, 36–41.

Reale, E., Spinello, A.O., Varinetti, E., & Zinilli, A (2023). *Documentation of RISIS datasets: EFIL (2.0)*. RISIS 2 - European Research Infrastructure for Science, technology and Innovation policy Studies 2, EU-Horizon2020 Research and Innovation Programme Grant Agreement n. 824091. https://doi.org/10.5281/zenodo.7938134

Romero-Perdomo, F., Carvajalino-Umaña, J. D., Moreno-Gallego, J. L., Ardila, N., & González-Curbelo, M. Á. (2022). Research trends on climate change and circular economy from a knowledge mapping perspective. *Sustainability, 14*(1), 521.

Schneider, F., Buser, T., Keller, R., Tribaldos, T., & Rist, S. (2019). Research funding programmes aiming for societal transformations: Ten key stages. *Science and Public Policy, 46*(3), 463–478.

Spano, D., Mereu, V., Bacciu V., Marras S., Trabucco A., Adinolfi, M., ..., Zavatarelli M. (2020). Analisi del rischio. I cambiamenti climatici in Italia (Risk analysis. Climate change in Italy). Fondazione CMCC - Centro Euro-Mediterraneo sui Cambiamenti Climatici, 10.25424/cmcc/analisi_del_rischio

Spinello, A. O., Reale, E., & Zinilli, A. (2021). Outlining the orientation toward socially relevant issues in competitive R&D funding instruments. *Frontiers in Research Metrics and Analytics, 6*, Art. 712839.

Spinello, A.O., Varinetti, E., Zinilli, A., & Reale, E. (2023). Il finanziamento competitivo della Ricerca e Sviluppo in Italia per le sfide sociali e tecnologiche (Competitive R&D funding in Italy for social and technological challenges). In Consiglio Nazionale delle Ricerche (ed.), *Relazione sulla ricerca e l'innovazione in Italia. Analisi e dati di politica della scienza e della tecnologia. Quarta edizione (Report on Research and Innovation in Italy. Analysis and data on Science and technology policy* (4th edn, pp. 23–58). CNR Edizioni.

Westerhoff, L. (2010). 'Planning for today': The nature and emergence of adaptation measures in Italy. In E. C. H. Keskitalo (Ed.), *Developing adaptation policy and practice in Europe: Multi-level governance of climate change* (pp. 233–270). Springer.

Witte, J. (2023). Higher education, science and the climate crisis. In S. Marginson, B. Cantwell, D. Platonova, & A. Smolentseva (Eds.), *Assessing the contributions of higher education. Knowledge for a disordered world* (pp. 89–110). Edward Elgar Publishing.

Zahradnik, G., Heller-Schuh, B., & Scherngell, T. (2021). *Conceptual framework: EUPRO deepening (NATPRO)*. RISIS 2 - European Research Infrastructure for Science, technology and Innovation policy Studies 2, EU-Horizon2020 Research and Innovation Programme Grant Agreement n. 824091. https://doi.org/10.5281/zenodo.4610742

CHAPTER 5

UK Universities and Climate Change

David Charles

Introduction

A concern for climate change in UK universities dates back to the formation of the Climatic Research Unit (CRU) at the University of East Anglia in 1971. The CRU and other later research groups played a critical role in the identification of climate change as a phenomenon which could be influenced in the short term through the effects of human action. These initial scientific efforts contributed to the first Intergovernmental Panel on Climate Change (IPCC) report in 1990. The decade that followed saw considerable investment in climate change-related research by the UK government through UK universities and constituted a period of mainly research and the nascent development of policy. The new century saw a shift in response by UK universities to the climate agenda. Whilst research on detecting climate change increased, there was a wider re-orientation towards solutions to the problem, an increased focus on teaching, and the start of actions by universities themselves to reduce their carbon emissions. The Climate Change Act of 2008 required universities to respond with

D. Charles (✉)
Newcastle Business School, Northumbria University,
Newcastle upon Tyne, England
e-mail: david.charles@northumbria.ac.uk

J. Kekäle, R. Pinheiro (eds.), *Higher Education Policy for Tackling Climate Change*, https://doi.org/10.1007/978-3-031-78479-8_5

the development of strategies for climate change mitigation and a requirement to reduce, measure, review, and report on emissions. Subsequently, as the agenda has further developed and become more urgent, climate change strategies have become more strongly embedded in universities, with multi-dimensional aspects.

This chapter outlines the emergence and growth of UK university strategies to combat climate change and focuses on several specific aspects of the current university policy environment. A major theme is the collaboration across the sector, with a number of networks and groups being formed to share knowledge about decarbonisation and the ways in which universities can contribute to combatting climate change. This includes the UK Universities Climate Network and collaboration around the CoP26 meeting which was held in the UK. Examples of actions from several UK universities will be provided, particularly from the North East of England. Another central issue is the direct impact of universities' operations in terms of emissions and the requirement to reduce carbon-based energy is a high priority across UK universities. The chapter will provide examples of the measures being adopted and where the experiences of decarbonisation are being shared with local partners in local energy networks. More challenging is the impact of travel, with UK universities being heavily committed to internationalisation of education and research collaboration, both of which involve considerable international air travel. Progress is less well developed in this area, with universities turning a blind eye to the travel of students to the UK to study.

Overall, the chapter seeks to provide an overview of how UK universities have committed to the climate change objective, including their role as engaged or civic universities with a responsibility to take a leadership role within their communities. However, the successful implementation of actions to reduce climate change gas emissions and to assist the wider community of firms and institutions in mitigation and adaptation also depends on their partners, and especially the support from the central government. As will be shown, this is stronger in some aspects than others.

A Framework for Examining Sustainability and Universities

Climate change has a broad impact on the university mission, affecting all aspects of the university's activities and offering huge opportunities as well as costs and responsibilities (M'Gonigle & Starke, 2006; Rappaport &

Creighton, 2007). In both of the core missions of teaching and research, sustainability has been one of the fastest growing themes in recent decades as universities have taken opportunities, stimulated by public research funders, to create new research institutes and departments, as well as to embed aspects of sustainability in a wide range of existing disciplines, from life sciences to engineering, geography to business and economics, and even in aspects of the humanities. UK universities have been responsive to the opportunities arising from cross-disciplinary and policy-engaged work, often identified with the idea of mode 2 knowledge production (Gibbons et al., 1994). Research funding has been generally supportive of interdisciplinarity, sometimes driving change, although sometimes falling back on disciplinary silos, but universities have generally recognised interdisciplinary 'grand challenges' in a positive way, often embedding them in their institutional strategies. New research institutes and interdisciplinary teaching programmes have been supported, with climate change-related topics often to the fore. Sustainability is also central to the emerging notion of the engaged or civic university (Civic University Network, 2022; Goddard & Vallance, 2013), in which universities take on responsibilities within civic society, both locally and globally.

There are four ways in which HEIs contribute to sustainable development, reflecting the diverse roles of universities as large institutions, sources of expertise, and providers of common educational experiences.

The first is in the universities' own *institutional management practices* as institutions have developed their own sustainable development policies, statements and visions, often supported by a Sustainability or Environmental Policy Group. From this, many universities have developed sustainable policies; estates management practices are central, and some of the more innovative approaches taken have embodied a sustainable systems approach to estates management (Simkins & Nolan, 2004, and see, e.g. the EcoCampus accreditation https://ecocampus.uk/). New buildings have used the latest technologies and techniques (often developed within the university themselves) to minimise energy use, suppliers have been audited for their own sustainable credentials, and the practices of the universities have been redesigned to minimise energy and resource use. One of the greatest impacts of universities on their environment is their travel footprint, particularly for campus universities in out of town locations, and many have attempted to reduce travel by staff and students by levying parking charges whilst subsidising public transport, car sharing, and home working arrangements (Logan et al., 2020). However, this presents particular challenges for internationalisation, with increasing numbers of international students (Universities UK International, 2022). Some universities have also changed their own estate management practices to encompass traditional techniques on their farms and woodlands to encourage biodiversity and support nature conservation efforts (see, e.g. https://www.lboro.ac.uk/services/sustainability/biodiversity/).

Secondly, universities themselves are *sources of research and expertise* in the technical and holistic aspects of sustainable development. In terms of policy development, universities have been central to the creation of a knowledge base which has informed government policies towards climate change, both from the point of view of the underlying science and in terms of mitigation and adaptation technologies. Universities that have drawn on their own expertise in campus development implicitly recognise this asset, and many universities have sought to commercialise this expertise. Many firms are currently introducing environmental management systems (EMSs) (Simkins & Nolan, 2004), and a range of universities offer consultancy services to facilitate and embed the introduction of these systems to maximise business competitiveness. Research activities are extremely diverse, from the commercially sensitive development of new EMS techniques to much more socially embedded research projects; sea-level rise and flooding are two environmental issues with profound social implications at a local level, and a number of universities have used their social science expertise to assess the economics underpinning land use and flood defence issues at a local level (Fankhauser et al., 2019; Hulme & Turnpenny, 2004; Wreford et al., 2019).

Thirdly, universities have an extremely *important educational and cultural role* in providing a common educational experience for students, enculturating a set of social beliefs through academic practice. Sustainability is increasingly being formally incorporated into curricula, and steps are being taken in many institutions to ensure that all their students at least deal with and have an awareness of the subject as a prerequisite for graduation, much as graduation currently has independent, self-directed study as an award prerequisite. The educational process is also being actively consumed by a range of socially excluded individuals and groups, and through individuals' own learning and research, universities contribute to the social justice aims of sustainable development by supporting the development of a cadre of community leaders who are able to articulate and justify the needs of their own communities in the broader context of the sustainable development of UK society (Ralph & Stubbs, 2014).

Finally, universities have involved themselves with *institutions and activities* directed towards local and regional sustainability. A number of universities work very closely with local authorities and other sub-national bodies to ensure that their plans and policies are contributing to wider social and physical regeneration efforts. Universities have also become involved in a number of regional-scale activities seeking to design sustainability into regional planning activities (Charles & Benneworth, 2001; Leal Filho et al., 2023).

The Evolution of the Policy Environment in the UK

UK governments have typically had a hands-off relationship with the content of academic activity whether research or teaching, providing funding at varying levels over time through intermediary organisations. Universities are independent bodies, usually established by the act of Parliament, but with the government having no direct role in their management. Funding for research is provided through competitive grant schemes and a block grant related to quality, both now administered through UK Research and Innovation (UKRI), in which the so-called 'Haldane' principle (since 1918) insists that decisions on science funding are to be made by scientists through peer review and not politicians. Funding for teaching has also typically been allocated through an independent funding council, although mainly arising from student fees in which there are mainly no caps relating to numbers or disciplines. So the government influence over universities is mainly based on carrots (grants and incentives) or sermons on what universities should be doing (often driven through commissioned reviews). The use of direct and enforceable requirements (sticks) is very limited and mainly achieved through regulation and limits on funding. The government influence on the approach to climate change in universities has thus mainly been through carrots rather than sticks.

The emerging science of climate change in the 1980s prompted an early response by the UK government in a speech by the Prime Minister Margaret Thatcher to the Royal Society in 1988. Although initially somewhat sceptical of climate change, Thatcher (as a former chemist) had been convinced by the science and spoke out about the challenges of anthropogenic climate change, reduction of the ozone layer, and acid rain (Agar, 2019). This would stimulate considerable research and modelling, with UK-based research contributing to the signing of Agenda 21 and the Rio Declaration on Environment and Development at the UN Conference on

Environment and Development in 1992. The principles of sustainable development thus became increasingly central to a range of UK policies, including those focused on higher education. The 1993 Toyne Report, funded by the then Department of the Environment, recognised this and sketched out the progress necessary for HEIs to make a full contribution to the achievement of UK sustainable development (Toyne, 1993). In 1996, DoE funded a full review of the actions proposed in the Toyne Report, the Toyne Review, which noted that since 1993, little progress had been made and that the incentives for responsible and sustainable practice from HEIs were inadequately transparent (DoE, DfEE and Welsh Office, 1996).

Faced with an apparent weak response from the universities, the Department of Environment, Transport and the Regions (DETR) initiated a new study, the Higher Education 21 study, to sit alongside the Local Agenda 21 work being done by local government. This study, managed by the Forum for the Future, an NGO established by environmental campaigners, sought to more completely map the progress made by 25 selected institutions, termed 'trail blazers' towards various elements of sustainable practice (Forum for the Future, 1999). However, critical to this study was an examination of the barriers faced for the implementation of a good sustainability practice across a range of activities, from procurement and energy efficiency to greening the curriculum. The central finding of the study was that a strong institutional commitment to the principles of sustainability was essential if sustainable development was to be embedded into HEIs' practice. Case studies revealed university environmental policies, curriculum development, local partnerships, and some actions on energy use and procurement criteria, although subsequent data has shown that the impact of such work has been limited. A particular focus was placed on the need to develop local partnerships, such as with local government, around themes such as professional responsibility, global citizenship, regeneration, and research (Ali Khan et al., 1997).

After the conclusion of the HE21 project, universities took the findings and broadened their application beyond the particular project-based best practice examples and sought to deploy their assets to foster sustainable development, not merely institutionally but also the expertise of the university as an asset to assist others to move towards sustainable development. From 1997, there was an additional dimension to sustainability in England with the creation of the Regional Development Agencies and the emergence of a regional tier of strategic economic coordination, and consequently, sustainable regional development as a focus of their policy in this arena. All the English regions drafted regional sustainable development frameworks to provide strategic connectivity for a range of agencies and documents (Charles & Benneworth, 2001).

Climate Change Act

The emerging concern within the UK to reinforce a political response to climate change was underpinned in 2008 by the passing of the Climate Change Act (CCA). This was the world's first legally binding commitment to reduce greenhouse gas emissions, with a target of an 80% reduction on 1990 levels by 2050 (Lockwood, 2013). The Act received widespread political and public support and committed the UK to a series of carbon budgets proposed by an independent committee, to which the government needed to present policies for their achievement. In this way, the government was proposing a stick to be used against themselves as well as other public bodies to ensure a response to the climate challenge. The stick was something of a virtual stick though in that it was expected there would be compliance and there were no actual sanctions.

As part of this legislation universities and colleges, directly or indirectly, were required to reduce their carbon emissions to help reach the national target, but also to develop plans to reduce emissions, measure and review progress, and report to the public. These obligations were passed down by the relevant funding bodies for higher education, with sector-level targets. There was no system of incentives or penalties for this. The legislation simply makes it a requirement for public bodies to meet the target and to report their progress towards it. Universities have been more than willing to adopt the targets, and compliance with reporting has generally been good—often, universities have set themselves more demanding targets.

The government has also taken a hands-off position on how public bodies should meet their targets, and it is left to each sector to develop appropriate responses, usually through some form of sectoral committee or association. For higher education, there has been no real advice or any additional funding; indeed, recent years have seen the freezing of the undergraduate home fees which has meant an effective cut in unit resources with universities having to seek alternative forms of income such as international students. There has been no additional funding for meeting climate change mitigation costs.

In 2019, the UK committed to reducing greenhouse gas emissions to net zero by 2050 through a statutory instrument to amend the target of the CCA (2008 [2050 Target Amendment] Order 2019). This 'net zero' target was proposed by the Climate Change Committee and has become the first legally binding commitment of any country to reach net zero. Subsequently, the government has presented a strategic framework for progress to net zero (BEIS, 2021), although recently, there has been some backtracking by the current Sunak administration. Universities, however, have tended to bring forward their net zero targets. Leeds University, for example, has committed to net zero emissions by 2030. Working from a 2020 baseline for core campus-based emissions of 71,546 tCO2e, the university is making £150m in campus investments which will reduce this by 48,101 tCO2e by 2030, leaving 23,445 tCO2e that will need to be offset (University of Leeds, 2021).

All universities have adopted the net zero targets, although with varying degrees of progress so far, but with a range of different factors to consider—size of the institution, age of the physical estate, density of campus, previous investments in energy efficiency, research orientation, disciplinary mix, commuting patterns of staff and students, internationalisation, and managerial effectiveness.

One possible indicator of commitment to emissions reduction has been the declaration of a climate emergency, with Bristol leading the way in April 2019 and 36 other universities following over the following year (Latter & Capstick, 2021). These 37 are a cross-section of institutions. The first three were Russell Group (Bristol, Newcastle and Glasgow), but overall less than half of the Russell Group were in that first 37, and there were many post-92 universities, including some of the most recent designations (Falmouth, Canterbury Christ Church, Winchester, Plymouth Marjon). It included some of the largest institutions (UCL, Manchester, Cambridge) and smaller ones (Marjon and Royal Agricultural University). Latter and Capstick (2021) note though, that these declarations are in the first instance promotional statements, highlighting the work of the university, claiming leadership, and often timed to attract publicity. Some are undoubtedly very committed to addressing climate change, but as an indicator, it is imprecise, and we cannot know if some that did not declare a climate emergency at an early stage are not more committed to action than some that did.

The UK Universities Climate Network was 'founded in 2019 as the COP26 Universities Network with a goal to ensure the academic sector played its role in delivering a successful UN COP26 Climate Change Conference in Glasgow in November 2021' https://uucn.ac.uk/about-us/. The network undertook an extensive work programme in the lead-up to the conference to create momentum, engage public and business audiences, and provide support to the UK government. This then became the UK Universities Climate Network (UUCN) in March 2022 in order to continue their work to contribute to the creation of a resilient, net zero world. The network continues as a knowledge-sharing resource, publishing briefings and case studies on all aspects of climate change and, particularly, the actions that universities have been taking.

University Actions on Climate Change

The following sections set out some of the actions of UK universities in seeking to address the challenges of climate change and in pursuit of net zero emissions.

Campus Operations

All UK universities have been required since the CCA of 2008 to commit to the reduction of carbon emissions and to report on their progress. Measuring those carbon emissions is complex though and is defined through three levels.

- Scope 1 emissions consist of greenhouse gas (GHG) emissions that are directly made by the operations of the university, primarily through heating with fossil fuels and running vehicles such as service vans. This is relatively straightforward to address through, for example, the replacement of gas heating and the decarbonisation of the university vehicle fleet.
- Scope 2 emissions are indirect through the purchase of energy from other suppliers in the form of electricity generated by fossil fuels. Here, it is possible to buy electricity generated by renewable sources, and in the UK in 2022, around half of electricity is generated from renewable and nuclear sources (actually more as this excludes imports, mainly from French nuclear power).
- Scope 3 emissions are the hardest to deal with as these are emissions that the university is only indirectly responsible for through its supply chain and travel associated with staff and students. This set of emissions is still largely to be addressed.

UK universities have made positive moves in setting targets for net zero carbon emissions, with many setting target dates well in advance of the government goal of net zero by 2050. UK universities report as follows:

- '96% of our member universities have a clear, publicly available strategy to reduce carbon emissions;
- 97% have sustainability information openly available on their website;
- 82% report their carbon reduction figures, allowing the public to monitor the progress universities have made so far and assess their future progress' (Universities UK, 2022).

The area of least progress, though, is in scope 3 emissions, where measurement is difficult, and so far, only 6% of universities have committed to reductions. A recent study estimated that out of a total investment of £37 billion by universities needed to meet net zero challenges, £24.4 billion would be needed for supply chain issues, compared with £6.5 billion for the built environment and £5 billion for transport (BUFDG et al., 2023). The Royal Anniversary Trust (2023) study indicates that current expectations are for a 72% reduction in all emissions by 2040, but with continuing problems over supply chains, flights, and student accommodation, some of which would need to be offset to reach net zero.

In Newcastle, for example, the two universities have pursued different strategies for carbon emission reduction. Northumbria University has a slightly less compact campus, split by a major motorway, which has gradually expanded through acquisition, and so it has had a distributed heat system, previously with gas boilers for each building, although with a long-term commitment to photo-voltaic (PV) generation. The university was the first to put PV panels on the external walls of an existing building in the 1990s, and it has installed considerable capacity since. Currently, the university is also installing air-source heat pumps to replace gas boilers. Its current targets are an 80% reduction from 2015/16 by 2030 and net zero by 2040.

Newcastle University, with a more compact campus, previously had a central boiler for the campus core, although this has now been replaced with more distributed heating. The university is now looking to accelerate its target from a 2040 date for net zero by 2030, and is looking to new schemes such as rooftop solar panels. The current focus is on deep reductions in energy use.

Both universities are also currently in discussions with Newcastle City Council over a district heating scheme, as the two campuses sit on either side of the main council office building with the main central hospital adjacent also. This makes a very large single campus wrapping round the northern and eastern edge of the city centre, ideal for a district heating scheme. The focus in both cases is mostly on the built environment.

Travel

It was noted earlier under campus operations that scope 3 emissions include emissions related to travel, a difficult topic for universities on three grounds: the need for staff and students to commute to the campus, the dependency that UK universities have on international students and the associated long haul air travel, and the need for staff to attend international conferences and do international fieldwork (Hoolohan et al., 2021; Shields, 2019). Universities have largely been left to their own devices in developing strategies for decarbonising travel with no national guidance.

Green travel plans have been developed and implemented over many years, as is typical within the UK public sector. These focus primarily on local travel to campus and use a carrot-and-stick approach, and the stick mainly being the reduction of and increasing cost placed on car parking on campus. In the case of urban universities, parking has often been reduced through construction, whilst most universities now charge for parking through an annual permit which are usually limited in number. Incentives to use public transport or cycling can include savings on transport passes and schemes to provide reduced-cost bicycles (Logan et al., 2020). However, since the pandemic, there have been contradictory trends with more working from home to reduce travel as well as a switch from public transport to car use because of health concerns and reductions in service frequencies by public operators (Anable et al., 2022).

By comparison, little effort has been placed on international travel, apart from exhortation to reduce travel if possible. The complete ban on travel during the pandemic and the switch of much conference activity to online forms demonstrated the possibility of reduced conference travel, albeit with some loss of quality of experience (Zeferina & Hoolahan, 2022). The subsequent slow return to normal international travel has been accompanied by a greater test on the need for travel,[1] but ultimately

[1] See https://www.reading.ac.uk/procurement/insurance/business-travel/policies (accessed 3/4/2024) as just one example, as most universities request staff to think of the environment when planning travel.

there seems to be largely a return to business as usual for international academic conferences.

The biggest challenge for the UK is the travel associated with international students, often long haul from China or India. Student international travel accounts for 12% of total emissions from UK universities (Royal Anniversary Trust, 2023) and the BUFDG (2023) study assumes that the scale of these emissions cannot be abated by 2050 and will need to be offset.

Research Endeavours

UK universities have played a key role in the development of research on CO_2 and climate change, notably with the case of the University of East Anglia. The UEA in 2022 celebrated 50 years of research on climate change since the formation of the Climatic Research Unit in 1972. The CRU was established to research the past evolution of climate, providing data which established the extent of global warming in recent decades. With the establishment of the IPCC in 1988, UK universities played a strong role, contributing through their research reflecting the strength of UK research.

The UEA also led the formation of the Tyndall Centre for Climate Change Research in 2000 as a multi-site, multi-disciplinary research organisation to 'research, assess and communicate from a distinct trans-disciplinary perspective, the options to mitigate, and the necessities to adapt to current climate change and continuing global warming, and to integrate these into the global, UK and local contexts of sustainable development' (Fudan Tyndall Centre, n.d.). It brings together eight UK universities (University of East Anglia, University of Cambridge, Cardiff University, University of Manchester, Newcastle University, University of Oxford, University of Southampton, and the University of Sussex), each of which now houses its own Tyndall Centre, and Fudan University from China. Although this is a UK-funded network, it was thought necessary to have an international partner to ensure relevance for international policy. The Tyndall Centre had been developing links with several Chinese universities and selected Fudan as a core partner, especially as they were able to obtain financial support from the Chinese and Shanghai governments (£20 million for the first four years) (Contestabile, 2011).

The centre was named after John Tyndall a nineteenth-century Irish scientist who first demonstrated the greenhouse effect. The centre has been funded by several of the UK research councils as well as through one-off research grants. The centre has four main research themes: accelerating social transitions, building resilience, overcoming poverty with climate actions, and reaching zero emissions.

Throughout this period, the Natural Environment Research Council (NERC) has supported and invested in climate change research across UK universities. A particular focus has been on the development of environmental measurement systems that have been developed since the 1930s, including the work done through NERC institutes such as the British Antarctic Survey (BAS), and the National Centre for Atmospheric Science (NCAS). Whilst some of NERC's own centres have their own labs, they also have teams embedded in universities. The BAS was established in 1962, although the UK had operated research stations in Antarctica since 1944. The main offices are in Cambridge, with several stations in the Antarctic region examining climate change, ice-shelf changes, the ozone hole, ecological changes, sea-level rise, and so on. NCAS was more recently established in 2002 to coordinate atmospheric research across the UK with the involvement of 11 universities, but also working closely with the UK Meteorological Office and Environment Agency, and the US National Centre for Atmospheric Research as well as other international partners in Europe and Africa (https://ncas.ac.uk/about/history/).

Whilst a majority perhaps of NERC funding is related to climate change issues, there are significant investments from other research councils as well, notably for the Engineering and Physical Sciences (EPSRC), Economic and Social Sciences (ESRC) and even Arts and Humanities (AHRC). A report commissioned by ESRC on the social science of climate change found that between 2008 and spring 2019, UK research councils awarded grants to 481 climate change projects with strong social science components with an overall spend of £438 million (Fankhauser et al., 2019). Obviously, this is only a minority of all research council climate change projects, as many do not include a social science component. A current example of interdisciplinary working, however, is a small £11.5 million call on 'Resilient UK coastal communities and seas': this programme is co-funded by ESRC, AHRC, NERC, and the Department for Environment, Food and Rural Affairs, and focuses on understanding the impact of climate change on Britain's coasts, but also on the interactions with coastal communities, their health, economic, social and cultural challenges and potential interventions.

Teaching

The introduction of climate change issues into curriculum design is an important element of the university role in which there has been some criticism of slow progress (QAA/Advance HE, 2021). Assessment of progress needs to acknowledge three main levels of potential action. First, there are new programmes relating to climate change science, which may be linked to the emergence of major research centres and the transformation of disciplines. Second, there are specific modules which are developed for a wide range of disciplinary programmes, whether in environmental-related programmes such as geography, planning, or engineering, or in general disciplines such as business. Finally, there is the embedding of climate change material in existing modules or modules that are primarily focused on other issues.

Most universities have been active in developing dedicated sustainability and climate change-related programmes where they have the appropriate disciplines and have also developed a wide range of related modules across other disciplines. Taking examples from Northumbria University, at a programme level we can find for example a BSc in Environmental Sciences which has considerable content around different aspects of environmental change including climate change, or there is the MSc in Renewable and Sustainable Energy Technologies, or an MSc in Economics and Sustainability. All of these programmes are targeted at students with an interest in climate change and specific skills—environmental, engineering or economic—related to that topic. On a modular level, the geography programme has a module on 'Green cities and nature-based solutions', whilst all of the business and management programmes have a core first-year module on Business Economy and Society which focuses on sustainable development goals. Within modules, it is more difficult to identify climate change elements, and this is for individual staff to decide, but there are undoubtedly many examples.

The development of the curriculum in the UK is largely driven by departments and individual academic staff, although working within the context of university quality systems and recognising market demand via university marketing services. The HE21 project identified a growing number of curriculum developments in the late 1990s that were concerned with climate change, but these have continued to grow on an ad hoc basis in the absence of top-down guidelines. Government has no real input to the mix of programmes or their content, apart from a few regulated subjects such as medicine and teacher training. There is no centralised planning of discipline-level provision, especially since number caps on universities were removed a few years ago. However, the quality systems in most universities do require programmes and modules to take into account national benchmarks for disciplines, mapping the new programme or module against the benchmarks.

The Higher Education Academy (Advance HE), a membership body which supports teaching and learning practice, among other aspects of HE, and oversees professional standards, was active between 2010 and 2015 in funding pilot projects and developing materials for sustainability in the curriculum (QAA/Advance HE, 2021; Cross & Congreave, 2021). The other key national body, the Quality Assurance Agency, oversees guidance or benchmarks for disciplinary programmes, and prior to 2021, most disciplines made little or no mention of sustainability in their benchmarks unless it was core to the discipline, such as in geography or environmental studies. Consequently, the response of universities has been patchy. However, in March 2021, the Quality Assurance Agency and Advance HE produced a guidance document, 'Education for Sustainable Development Guidance' (QAA/Advance HE, 2021).

Subsequently, all QAA subject benchmark statements have been revised with a standard requirement to include a section on sustainability. Every programme must, therefore, consider including some reflection on sustainability when validated or revalidated, according to the suggestions for that discipline. The Business and Management benchmark statement notes that:

> *'The study of Business and Management addresses environmental, social, governance and economic concerns with the aim of creating a better world. The discipline can contribute towards the achievement of the responsible and sustainable development as described by the United Nations Sustainable Development Goals, partly in terms of disciplinary content, and partly via disciplinary method (…). More specifically, Business and Management courses can draw on the UN Principles for Responsible Management Education (PRME) to develop the capabilities of students so that they become future generators of sustainable value for business and society as well as work towards a sustainable global economy. There is an expectation by some accrediting bodies that ethics, responsibility and sustainability (ERS) are integrated into the curriculum within Business Schools.'* (QAA, 2023, p. 7)

It is also noted that graduates should be able to demonstrate knowledge in 'ethics, responsibility and sustainability: managing responsibly and behaving ethically, particularly in relation to social, cultural, legal, economic and environmental issues' (QAA, 2023, p. 13). Other disciplines as diverse as mathematics and theology have similar requirements to consider sustainability as a theme on which disciplinary tools may be applied.

The evidence on teaching-related actions is thus that whilst much creative practice has been introduced, there has been a lack of national consistency in the absence of disciplinary guidelines. This has now been addressed within the sector and should be rolled out as programmes are revalidated over the coming years.

Local Institutional Engagement

Involvement in local sustainability partnerships has developed from the adoption of Local Agenda 21 groups by local authorities in the late 1990s, which sought to bring together key local partners, including universities. This was followed up in the early 2000s by a central government requirement for each region to develop a framework for the implementation of sustainable development. In the Northeast, for example, a regional sustainable development round table was established, initially supported by the regional office of central government and regional development agency, but three of the local universities took on the responsibility for providing the secretariat, bringing together expert comment and advice (Charles & Benneworth, 2001). This regional approach to sustainability during the 2000s was, however, dismantled in the English regions after 2010 when the Conservative-led government removed the regional development agencies and regional government offices, leaving a governance gap at the regional scale. Subsequently, governance has been re-assembled piecemeal as new combined authorities have been established, bringing together city regions in some areas only (Newman et al., 2024).

Since 2019, the commitment of universities to net zero partnerships has been enhanced in several cases by the development of civic university agreements. These agreements were initiated by the report from a national Civic Universities Commission (UPP, 2019) launched by the UPP Foundation, a charitable foundation established by a provider of student accommodation. The report examined the engagement of universities in their civic contexts and made a recommendation that universities wishing to be seen as 'civic' should sign a civic university agreement (CUA) with local partners. These CUAs have been developed since 2020 and have been affected in some cases by the lockdowns and impact of the COVID-19 pandemic, with some in the larger cities particularly bringing together two or more universities in partnership. Net zero objectives have been written into several of these agreements.

In Greater Manchester, the CUA brings together five higher education institutions (Manchester, Bolton, Manchester Metropolitan, Salford, and Royal Northern College of Music) with the Greater Manchester Combined Authority (a combined authority working with the ten local authorities in the conurbation, and with devolved responsibilities from central government). The CUA was signed in 2021, partly motivated by the intense collaboration between universities and the GMCA during the pandemic, and has six priority themes connected to the wider Manchester strategic frameworks for economic development, health, inequality, and so on. One of these themes is net zero, and under this theme, the partners, working with the private sector, have established the Energy Innovation Agency as a city-wide programme to transform Manchester's energy systems, focusing on decarbonisation as well as the exploitation of new clean energy technologies.

Similar initiatives have been developed in other cities. In Nottingham, the two universities have an agreement with their local enterprise partnership, city council, county council, health organisations, and a vocational college. Again, sustainability is a key theme for proposed action focusing on collaborative research and translation into practice, implementing low-carbon infrastructure and developing local behavioural campaigns.

In Newcastle, the CUA, Collaborative Newcastle, brings together the two universities, city council, combined authority, NHS, and two third sector organisations. It was quickly decided in 2019 that a joint approach between the two universities was needed, although the final agreement was not signed for some time. As with Manchester, though the universities worked closely with city partners during the COVID-19 pandemic, focusing on battling the pandemic and on post-pandemic economic recovery (Charles, 2023). Here, the initial themes were people, planet and prosperity, with the planet theme focused on net zero. A series of collaborative projects was identified by the universities under these themes. In the case of net zero, the main focus was to share experiences over decarbonisation strategies and to develop a new heat network across the city.

Also, from 2022, the central government has established five Local Net Zero Hubs in England to bring together local and national governments in coordinating responses at the local level and oversee nationally funded projects. These hubs work through local enterprise partnerships with more local initiatives such as Net Zero North East. This latter body is a partnership of local authorities, business, public sector, education, and civil society to develop a regional approach to net zero. The universities in the region are represented on the board through a PVC from Newcastle University and contribute to developing the evidence base and specific projects.

Over time, universities in the UK have shown a strong willingness to take a leadership role within their regions as civic universities supporting partnerships and developing collaborative projects; however, this also depends on the regional governance frameworks that have been imposed by the central government, and which have been subject to considerable turbulence in recent years, oscillating between regional, local, and national scales.

Conclusions and the Way Forward

Whilst universities in the UK have been active in the climate change agenda for several decades in some respects, their involvement has not always been consistent across all the fields of activity which could be undertaken. The greatest level of activity has been in the research field, where the UK government has provided considerable encouragement and funding for climate change-related research, and UK universities have played an important international role in the development of knowledge. Direct action to reduce emissions from university-related activity has been much slower to develop, although most universities now have challenging targets for net zero emissions. Some progress is being made on this, but with a main focus on energy use, and some areas of scope 3 activity, such as travel, remain challenging, and policies have been slow to emerge.

Climate change education has lacked a core impetus until recently, with individual institutions and even departments and academic staff having to take on the responsibility of generating new curricula focused on climate change and sustainable behaviour. However, recent adoption of new benchmark standards should ensure a more holistic implementation of curriculum development across all disciplines rather than just the core climate science topics. In this, the government has been relatively inactive, and the action has been led within the HE sector itself.

The greatest variation perhaps has been in the area of local institutional engagement, where national government approaches towards local governance of climate policy have been fragmented and inconsistent, varying over time, and hence, the university role differs between regions and is voluntary.

The overall picture is positive, and many universities have declared a climate emergency, showing their commitment, but implementation has depended on the direction and support given by the national government: strong on research and weak on regional partnership development. The slow adoption of curriculum development reflects the emphasis on autonomy in the UK system, but the sector is now developing a more holistic approach. The difficult challenge now is to meet the targets for net zero emissions, an area where central government has set a target but has done little to help the sector achieve it. Indeed, the importance of internationalisation in the sector, both in terms of research collaboration and international student fees, will make it difficult to cut into the travel emissions. Government unwillingness to raise domestic fees or increase its own contribution leaves universities highly dependent on international student revenue and an emissions burden which is difficult to mitigate.

There is therefore a need for a more holistic approach to university policy in the UK if the net zero targets are to be achieved. Central to this is the development of an understanding of how various university policies interact around the climate change agenda, something which the government has not addressed, in considering all university policies in isolation, often sending conflicting signals—such as requiring more international students to offset falling UK fees, whilst making it more difficult for international students to enter the UK. Universities have been required to reduce emissions to reach targets set by the CCA but with a falling unit of resources and no dedicated funding for carbon reduction strategies. Whilst universities and NGOs are developing action plans for meeting net zero targets, the possibility of success is dependent on a more coordinated national approach than has been seen in recent years.

References

Agar, J. (2019). *Science policy under Thatcher*. UCL Press.

Ali Khan, S., Agyeman, J., & Gibson, M. (1997). *Higher education 21 local agenda 21: Partnerships for sustainability*. HE21/Forum for the Future.

Anable, J., Brown, L., Docherty, I., & Marsden, G. (2022). *Less is more: Changing travel in a post-pandemic society*. Oxford Centre for Research into Energy Demand Solutions.

British Universities Finance Directors Group, Environmental Association for Universities and Colleges, and Association of University Directors of Estates. (2023). *The cost of Net Zero: Outcomes from a project exploring the costs of Net Zero for Higher and Further Education in the United Kingdom*. BUFDG/EAUC/AUDE. https://www.eauc.org.uk/file_uploads/20230524_hfe_v4_0_-_cost_of_net_zero_report_1.pdf

Charles, D. (2023). University-civic engagement in the time of the pandemic. In R. Pinheiro, E. Balbachevsky, P. Pillay, & A. Yonezawa (Eds.), *The impact of Covid-19 on the institutional fabric of higher education: Old patterns, new dynamics, and changing rules?* (pp. 223–244). Cham.

Charles, D. R., & Benneworth, P. (2001). *The regional mission: The regional contribution of higher education: National report*. Universities UK.

Civic University Network. (2022). *The role of HEIs in the climate action agenda Supporting place and communities, in a just transition to net zero*. Civic University Network. https://civicuniversitynetwork.co.uk/wp-content/uploads/2022/06/Report-The-role-of-HEIs-in-the-climate-action-agenda..pdf

Contestabile, M. (2011). Partnering with China. *Nature Climate Change, 1*, 320. https://doi.org/10.1038/nclimate1207

Cross, I. D., & Congreave, A. (2021). Teaching (super) wicked problems: Authentic learning about climate change. *Journal of Geography in Higher Education, 45*(4), 491–516. https://doi.org/10.1080/03098265.2020.1849066

Department for Business, Energy and Industrial Strategy (BEIS). (2021). *Net zero strategy: Build back greener*. HMSO.

DoE, D. E. E., & Office, W. (1996). *Environmental responsibility: A review of the 1993 Toyne report*. The Stationery Office.

Fankhauser, S., de Menezes, A., & Opacic, N. (2019). *UK research on the social science of climate change. A synthesis of ESRC and related investments*. Policy Report. Place-Based Climate Action Network, London School of Economics and Political Science.

Forum for the Future. (1999). *Learning for life: Higher education and sustainable development*. Higher Education 21, Forum for the Future.

Fudan Tyndall Centre. (n.d.). *Fudan Tyndall Centre*. http://tyndallcentre.fudan.edu.cn/en/show.aspx?info_lb=595&flag=538&info_id=2457

Gibbons, M., Limoges, C., Nowotny, H., Schwartzman, S., Scott, P., & Trow, M. (1994). *The new production of knowledge: The dynamics of science and research in contemporary societies*. Sage.

Goddard, J., & Vallance, P. (2013). *The university and the city*. Routledge.

Hoolohan, C., McLachlan, C., Jones, C., Larkin, A., Birch, C., Mander, S., & Broderick, J. (2021). Responding to the climate emergency: How are UK universities establishing sustainable workplace routines for flying and food? *Climate Policy, 21*(7), 853–867. https://doi.org/10.1080/14693062.2021.1881426

Hulme, M., & Turnpenny, J. (2004). Understanding and managing climate change: The UK experience. *The Geographical Journal, 170*, 105–115.

Latter, B., & Capstick, S. (2021). Climate emergency: UK universities' declarations and their role in responding to climate change. *Frontiers in Sustainability, 2*. https://www.frontiersin.org/articles/10.3389/frsus.2021.660596

Leal Filho, W., Weissenberger, S., Luetz, J. M., Sierra, J., Rampasso, I. S., Sharifi, A., Anholon, R., Pires Eustachio, J. H. P., & Kovaleva, M. (2023). Towards a greater engagement of universities in addressing climate change challenges. *Scientific Reports, 13*, 19030. https://doi.org/10.1038/s41598-023-45866-x

Lockwood, M. (2013). The political sustainability of climate policy: The case of the UK Climate Change Act. *Global Environmental Change, 23*, 1339–1348.

Logan, K. G., Nelson, J. D., Osbeck, C., Chapman, J. D., & Hastings, A. (2020). The application of travel demand management initiatives within a university setting. *Case Studies on Transport Policy, 8*(4), 1426–1439.

M'Gonigle, M., & Starke, J. (2006). *Planet U: Sustaining the world, reinventing the university*. New Society Publishers.

Newman, J., Mukherjee, A., & Hoole, C. (2024). Constraints and enablers of regional environmental policy: Governance challenges in England and Wales. *Journal of Environmental Policy & Planning, 26*(1), 1–13. https://doi.org/10.1080/1523908X.2023.2297793

Quality Assurance Agency. (2023). *Subject benchmark statement: Business and management*. QAA. https://www.qaa.ac.uk/the-quality-code/subject-benchmark-statements/subject-benchmark-statement-business-and-management

Quality Assurance Agency for Higher Education and Advance HE. (2021). *Education for sustainable development guidance*. QAA.

Ralph, M., & Stubbs, W. (2014). Integrating environmental sustainability into universities. *Higher Education, 67*, 71–90. https://doi.org/10.1007/s10734-013-9641-9

Rappaport, A., & Creighton, S. H. (2007). *Degrees that matter: Climate change and the university*. MIT Press.

Royal Anniversary Trust. (2023). *Accelerating the UK Tertiary Sector towards Net Zero: A Sector-led Proposal for Action and Connected Thinking*. The Royal Anniversary Trust, London. https://royalanniversarytrust.org.uk/wp-content/uploads/2024/10/Accelerating-towards-Net-Zero.pdf

Shields, R. (2019). The sustainability of international higher education: Student mobility and global climate change. *Journal of Cleaner Production, 217*, 594–602.

Simkins, G., & Nolan, A. (2004). *Environmental management systems in universities*. Environmental Association for Universities and Colleges. https://www.sustainabilityexchange.ac.uk/files/emsiu-v5_1.pdf

Toyne, P. (1993). *Environmental responsibility: An agenda for further and higher education. DfE and Welsh Office*. The Stationery Office.

Universities UK. (2022). Climate crisis: What progress have universities made? *Insight and Analysis*, November 10. https://www.universitiesuk.ac.uk/latest/insights-and-analysis/climate-crisis-what-progress-have

Universities UK International. (2022). *How the Climate Crisis affects Internationalisation: Report of a survey into internationalisation and climate action*. UUK.

University of Leeds. (2021). *Pathway to net zero emissions by 2030*. https://www.leeds.ac.uk/download/downloads/id/2968/the-university-of-leeds-pathway-to-net-zero-emissions-by-2030.pdf

UPP Foundation Civic University Commission. (2019). *Truly Civic: Strengthening the connection between universities and their places*. UPP Foundation.

Wreford, A., Peace, S., Reed, M., Bandola-Gill, J., Low, R., & Cross, A. (2019). Evidence-informed climate policy: Mobilising strategic research and pooling expertise for rapid evidence generation. *Climatic Change, 156*, 171–190.

Zeferina, V., & Hoolahan, C. (2022). *Academic aeromobility post-COVID 19*. University of Manchester. https://pure.manchester.ac.uk/ws/portalfiles/portal/215957123/Academic_Aeromobility_Post_Covid.pdf

CHAPTER 6

Mapping Efforts in Polish Higher Education to Tackle Climate Change

Dominik Antonowicz, Michał Wróblewski, Joanna Suchomska, and Anna Budzanowska

Introduction

Polish universities are historically built on the Humboldt tradition (Dobbins, 2015; Nybom, 2003a) as social institutions that rarely and reluctantly engage vis-a-vis the external environment. Traditionally, they have been more concerned about academic freedom, institutional autonomy and self-governance than addressing social issues. However, the role of higher education (HE) institutions (HEIs) appears to be gradually changing, with a growing focus on their public mission, particularly considering the social

D. Antonowicz (✉)
Department of Science and Higher Education Studies, Nicolaus Copernicus University, Toruń, Poland
e-mail: Dominik.Antonowicz@umk.pl

M. Wróblewski
Institute of Sociology, Nicolaus Copernicus University, Toruń, Poland
e-mail: michwrob@umk.pl

J. Kekäle, R. Pinheiro (eds.), *Higher Education Policy for Tackling Climate Change*, https://doi.org/10.1007/978-3-031-78479-8_6

prominence of grand challenges, such as climate change (Kaldeway, 2018). According to UN documents, climate change is one of the biggest challenges humans have ever faced. Not only does it significantly damage the natural environment, but it also holds profound and far-reaching implications for a broad spectrum of issues, including food security, natural resources and migration patterns. Climate change is also a so-called 'crisis multiplier' that potentially poses catastrophic threats to peace and political stability worldwide (Dennig et al., 2015; Dercon & Christiaensen, 2011). Hence, this chapter aims to map the extent to which the problem of climate change has been recognized in Polish HE as *a grand challenge* (Kaldeway, 2018) and how (and by whom) it is being addressed.

The basic assumption of this chapter is that HE is a complex, multi-actor and multi-level phenomenon (Meng-Hsuan et al., 2017). It will hence be analysed on three major levels: (a) public policy (macro), (b) institutional (meso) and (c) individual academics (micro). We recognize that each level represents different types of actors, agendas and capacities, but they all make (and contribute to) the national system of HE. In short, the chapter presents systematic observations of structural tensions related to the HE sector's social responsibility for addressing grand challenges for humanity. This study makes two types of contributions—empirical, by drawing an empirical map of scattered initiatives designed (directly or remotely) across the three levels of Polish HE, and conceptual, by discussing the role of institutional logic as a major factor affecting institutional resilience to engage actively in addressing grand challenges.

Policy Dynamics of the Polish HE Sector

The HE sector in Poland consists of predominantly public universities and, to a lesser extent, other types of HEIs, such as private universities and the public University of Applied Sciences. The sector is formally still a

---

J. Suchomska
Department of Human Ecology at the Institute of Sociology, The Nicolaus Copernicus University, Toruń, Poland
e-mail: jsuchomska@doktorant.umk.pl

A. Budzanowska
Silesian University in Katowice, Katowice, Poland
e-mail: anna.budzanowska@us.edu.pl

binary system, though both types of institutions are regulated by the same legal act and are subject to a similar governing approach by the ministry. Historically, until very recently, Polish universities were regarded as a 'republic of science' (Polanyi, 1962), a 'republic of scholars' (Brubacher, 1967) and 'organised anarchies' (Cohen et al., 1972); they still stand out by the traditional model of institutional governance with a particularly broad scope of institutional autonomy and a strong notion of self-governance (Kwiek, 2015). Moreover, they are characterized by the steadfastness of historical institutions and a strong resistance against any instrumental or utilitarian approach applied to them (Antonowicz et al., 2020; Kwiek, 2015; Shaw, 2019). In principle, Polish universities see themselves primarily as social institutions of great cultural prominence and not as organizations (Maassen & Olsen, 2007) designed to address social and economic problems.

In the Polish HE sector, the government is generally a 'lone coordinator' (Huisman, 2009a, p. 2) that tries to steer HE. Its influence on universities is substantially limited by the universities' institutional autonomy and self-governing mode of governance (Dobbins, 2015; Kwiek, 2015). In addition, Poland represents a post-conflict society shaped by decades of communist rule that damaged the relationship between universities and the government (Białecki & Dąbrowa-Szefler, 1994; Trnavcevic, 2010). Since the democratic transformation (1989), Polish governments have been rather reserved in proposing policy agendas in the field of HE, affording academic institutions considerable autonomy to determine their performance. That said, despite the transnational economic and political drivers seeking to make HEIs more engaged in the development of a post-industrial economic model (Bell, 1976) and later the knowledge economy (Marginson, 2010), the Polish HE sector remains rather immune to those trends and mitigates their impact. So, Polish HEIs have only a symbolic and largely remote influence in shaping the national policy agenda (Cloete et al., 2006) and institutional strategies (Beerkens, 2008). This is largely due to Polish universities being driven by a conservative institutional logic (Thornton & Ocasio, 2008, p. 101) that assumes universities are exceptional institutions designed and built on Humboldtian principles, with an emphasis on performing traditional tasks of teaching and curiosity-driven research. Accordingly, contributing substantially to the wider society and economy does not seem to be a part of their institutional blueprint.

Historically, Polish universities were not designed to solve social and economic problems outside the ivory tower, and the alarming voices calling them were predominantly unanswered (Brzeziński & Nowak, 1997). However, as much as they are disconnected from the external environment, they are not immune to the reputational game in the HE field. On the contrary, their position in international rankings becomes not only the prevailing motive in policy discussions but also a central part of institutional strategies. The national government found in universities the fertile soil for the academic arms race (Enders, 2014) or academic star wars (Yudkevich et al., 2023) for the status of world-class universities, underpinned and driven by the focal concept of 'academic excellence' (Ramirez & Tiplic, 2014). Overall, the concept of academic excellence—a highly ambiguous notion—was narrowly operationalized through the global bibliographic databases as a measurable proxy of the quality of research (Cremonini et al., 2014). It builds its foundation for *seemingly* 'legitimate' transnational comparisons in the form of university league tables or rankings (Hazelkorn, 2011) that stimulate positional competition for exceptional status (Enders, 2014, p. 155).

The logic behind the reputational game is entirely internal and rests on the assumption that universities operate in a fairly closed field in which they are 'continuously striving to perform better than others' (Brankovic, 2022, p. 801). It impacts policy dynamics and often triggers proactive initiatives that specifically target the advancement in global rankings. They are particularly attractive for catching-up countries that lag in research and development expenditure, labour productivity and the visibility of research outcomes (Hazelkorn, 2011). The problem is that such an approach separates universities from their broader environment and the needs of broader society. Instead, it steers them towards mainly internal games for obtaining the limited resources of the prestige economy. Such a policy fits well into the historically developed institutional logic (and purpose) of Polish universities that particularly struggle with international visibility and reputation. Needless to say, Polish political leaders hope that universities perform a more central role in the distribution of global prestige and higher standings in the rankings. Unsurprisingly, Poland and several other countries have adopted policies, and Polish universities have developed strategies to maximize their reputation and status (Hazelkorn, 2014, p. 14). Unfortunately, it presents the HE sector as a pretty much closed system that primarily seeks to increase its institutional reputation through attempts to climb in the league tables. What is worse, it paints a false image of the purpose of HEIs as reputational vehicles, which can, in addition, serve only as a plausible tool for economic development. Both the historical legacy and the global reputational pressure create specific conditions in which addressing global challenges such as climate change is neither a policy priority nor a strategic agency for HEIs.

In short, the historically shaped institutional logic of Polish universities has a massive impact on their self-perceived role, designed to be immune to the needs of the broader society. Universities have been reluctant in committing themselves to address economic needs through applied research, knowledge transfer or professional education. However, both the government and universities have expressed considerable interest in academic excellence that shifts attention to the race for academic prestige. They point to their standings in different international league tables as policy and institutional goals as a source of reputation and institutional prestige. That being said, the goal of tackling grand challenges such as climate change only remotely corresponds with what matters in the academic star wars (Yudkevich et al., 2023). Moreover, striving for the status of a research university enables Polish universities to bail out from service to society at large. Of course, it does not mean that academics do not investigate social problems of paramount importance, produce new knowledge or seek new empirical findings. It only means that the Polish government does not exert political pressure on HEIs, and instead, the university management provides academics with nothing but freedom to study problems of their choice.

Methodology

HE, as a field of study, presents a multidimensional phenomenon that can be analysed in at least three independent but also connected levels: policy level (*macro*), institutional level (*meso*) and the level of academic profession (*micro*). In addition, HE is also externally influenced by transnational rules, norms and values (DiMaggio & Powell, 1983). Addressing wicked problems such as climate change requires considerable time and substantial resources (people, infrastructure and funds); it requires commitment and interplay among different levels and the involvement of multiple actors. In our study, we recognize that those levels are driven by different logic, but the actors play their roles, to a large extent, autonomously.

This study aims to map efforts undertaken in the HE sector to address climate change. Accordingly, we decided to examine each of the three levels separately because such efforts appear in certain social and political contexts that need to be considered while discussing them. Further, the study on each level of the HE system is based on different empirical data that require a distinctive methodological approach and different analytical toolkits. Last but not least, conducting a separate analysis on each system level helps to systematize the overall examination and frames the findings.

Studying policy initiatives aimed at facilitating the HE sector in tackling climate change is a challenging endeavour because they appear in different shapes across the public sector, primarily because 'policy initiative' is an ambiguous concept and can be manifested through different policy instruments (de Boer & van Vught, 2020), regulation, financial initiative or just a narrative. In addition, the policy initiatives involving HE may also come from different policy actors in the public domain. Therefore, we aim to identify (a) whether the populist Polish government (that stayed in power between 2016 and 2023) has committed to tackling climate change and (b) how this commitment has been translated into public policy. Thus, the analysis conducted on the policy level is less systematic and takes a snapshot of the policy realm, aiming to capture signals from the government and its agencies between 2015 and 2023.

The analysis of HEIs focuses on institutional strategies—publicly accessible documents through which a university tries to communicate (mostly internally) about its mission, goals and values. After 2011, public universities are mandated to prepare institutional strategies and use them as an opportunity to demonstrate their strategic agency (Piotrowska-Piątek, 2015). These key documents must be approved by the senates of public universities, which makes them a legitimate voice of the academic community. In Poland, there are approximately 120 public HEIs, most of which are regional and teaching-oriented. They employ around 85% of the academic staff and enrol approximately 70% of the students in the country (GUS, 2023). Besides 103 public universities, there are 138 non-public HEI and 32 public universities of applied sciences. Needless to say, a vast majority of HEIs (both public and private) are focused solely on education and professional training to address local labour market demands. It is unrealistic and perhaps also unreasonable to expect them to tackle major and 'wicked problems' such as grand (global) challenges. So, for the current study, we intentionally selected a sample of 20 public universities recognized for conducting both research and teaching and standing out by their links to the transnational field of global science. The sample of universities chosen for the study was selected based on the Polish Excellence Initiative (IDUB) which selected institutions with strong international research profiles. Ten scholars are laureates of IDUB and have received an additional 10% of public funding, and the others are runners-up (receiving an additional 2% of public funding). By no means are they representative of the entire Polish HE sector, but they stand for the research-intensive institutions that should be engaged in tackling global challenges. We conducted thematic reviews of each of the strategies from the selected HEIs, specifically considering five types of references: (a) climate change, (b) sustainable development, (c) ageing, (d) social inequalities and (e) social responsibility. Purposely, we took a broader view of the range of grand challenges to learn whether HEIs conceive themselves as committed to something else than academic excellence and the commercialization of knowledge.

The third layer of the analysis levels refers to the activities of individual academics. It is extremely difficult to measure individual interest and engagement in specific research. We believe the best available proxy for mapping academic interest is a list of projects awarded by the National Research Council (Narodowe Centrum Nauki—NCN). The NCN is an autonomous public agency that funds basic research through competitive schemes. In Poland, most, if not all, major research projects are funded by the NCN; therefore, it can be considered a solid measure of research interest. We analysed projects and other scientific activities funded by the NCN between 2016 and 2022. The criteria for selection of the analysed projects was their compliance with the provisions of Directive 2003/87/EC of the European Parliament and European Council of 13 October 2003 on establishing a scheme for greenhouse gas emission allowance trading within the European Union (EU).

The analysis conducted on three levels (shown below) provides empirical evidence for the overall assessment (mapping) of efforts undertaken in the HE sector to tackle climate change, which will help us determine the contribution of the Polish HE sector to addressing grand challenges like climate change.

Empirical Findings

The Level of Public Policy

HE policy is a complex phenomenon studied mostly because of continued and almost never-ending reforms (Antonowicz et al., 2020; de Boer et al., 2007; Trowler, 2002). For this study, we use a very generic definition of public policy as 'anything a government chooses to do or not to do' (Dye,

1972, p. 2) because governments influence the HE sector in various ways. The exceptional status of universities in Poland has considerable influence on policy instruments governments use to achieve their political goals, and those policy instruments offer a toolkit of solutions designed to prompt (or prevent) social change by public authorities (Capano & Howlett, 2020, p. 1). In this study, we examined the actions that the Polish government took to address the problem of climate change, focusing on those aimed at engaging the HE sector. To do so, we systematized our analysis by using Vedung's (1998) classical trichotomy of policy instruments: economic means (carrots), regulation (sticks) and information (sermons). The timeframe of the policy analysis covers the period 2015–2023. It begins with the Paris Agreement (2015), in which the Polish government committed to adopting a more serious approach to climate change, which one may expect to be reflected in public policy. The studies on policy discourse (Dziedziczak-Foltyn, 2017) leave little doubt that before this period, focus on the problem of climate change was non-existent in Polish public policy.

This study primarily focuses on the Ministry for Higher Education and Science, which acts as the so-called 'lone coordinator' (Huisman, 2009b, p. 2). We also try to provide inputs from other policy actors who can exercise indirect influence on the first stage of policy formulation—setting the agenda and problem definition (de Boer et al., 2017, p. 9). Typically, the Ministry identifies problems and announces its political agenda through sectorial strategies and programmes. We reviewed the strategic programme for HE 2015–2030 (MNiSW, 2015), which set four strategic goals, namely, (a) enhancing the quality of HE and its relevant social and economic needs, (b) increasing the quality of research, (c) strengthening institutional governance and finances, and (d) leveraging the impact of HE on their social, economic and international environment. The latter refers mostly to the economic aspect of research and development, the visibility of HEIs in international rankings and the development of life-long learning.

Overall, until 2022, the Ministry had not recognized the grand challenges as strategic goals for HE and thus did not have any specific policy in that regard. We also examined other less strategic documents (listed in the appendix) issued by the Ministry and other actors, finding hardly any acknowledgement of climate change. It paints a rather disappointing picture of climate change being considered as either an issue not worth special concern or something happening to some other people. More recently, the Ministry published another document, named the National Science

Policy (MEN, 2022, p. 36), which had signs suggesting that climate change should be an important point of reference for technology development in Poland, for example, in the context of new energy storage technologies, Renewable Energy Sources (RES) or transmission infrastructure development.

> *The primary objective of research should be to support the transformation of the energy sector, ensuring energy security, maintaining the competitiveness of the economy, energy efficiency reducing the environmental and climate impact of the energy sector.* (MEN, 2022, p. 36)

The role, status and mutual relationships of various sectorial strategies, strategic plans and long-term development programmes are more than ambiguous. However, these documents set the stage for policy design and, at least theoretically, demonstrate the guiding principles for agenda setting in developing HE policy. So, there has been a notable change in acknowledging the challenges posed by climate change, but it is depicted as a key economic challenge rather than a broader call for mobilizing the entire HE sector to be more engaged with the topic.

Sticks—Regulations

Overall, Polish HEIs enjoy a high degree of institutional autonomy concerning research and teaching. The Ministry (except in very specific programmes such as medicine, veterinary or nursing) does not have any instrument to influence universities about specific curricula requirements. Even during evident democratic backsliding (Grzymala-Busse, 2018), the institutional research agenda and curriculum development remained largely beyond direct political influence. However, more recently, the Ministry criticized university education for pursuing progressive ideological goals (e.g. gender studies) and wanted them to be more aligned with government policy. Needless to say, tackling climate change did not fit into the political agenda of right-wing populistic governments like the Law and Justice (Prawo i Sprawiedliwość—PiS). Further, there have been no other calls for introducing any additional HE regulations regarding tackling climate change, for example, in reducing carbon footprints among academics who travel more frequently than other professionals. No other policy actors (such as the Rectors Conference and the Students' Parliament) demonstrated a particular interest in tackling climate change because it has not been a policy issue in the public realm. Why such prominent challenges failed to show up in mainstream public debates is intriguing, but it should be noted that until 2022, Polish governments had been rather sceptic towards climate change. Concurrently, the opposition parties had different and perhaps more urgent political priorities, such as liberal democracy, an independent justice system and a free media. Most recently, the Russian invasion of Ukraine (2022) posed imminent and direct threats to Poland, pushing aside all other issues, including the grand challenges. However, Poland is mandated to implement the green policy due to intergovernmental agreements and, more specifically, membership of the EU. Poland has committed to introducing sustainable policy goals, the most important of which is a reduction in the use of coal in electricity production from 70% (in 2022) to 50% by 2030. However, some mid-term targets in these agreements were achieved through pure manipulations (for more on this, see Wiejski, 2022).

There are strategic documents that are not strictly focused on HE but can be indirectly related to HEIs concerning climate policy. Examples include Poland's Energy Policy 2040 (Ministry of Climate and Environment, 2021) and Industrial Policy (Ministry of Economic Development and Technology, 2021), which, in many passages, refer to technologies that are key due to the energy transition associated with tackling climate change—for example, nuclear energy and the hydrogen economy—which require R&D activities. Though some of them refer to the role of universities in these documents, they do not set clear goals and areas of responsibility for HEIs. Instead, they focus more on the role of the private sector and state-owned companies. A good example of this is the government's initiatives related to the development of the hydrogen economy. The key document initiating work in this area—the so-called Sectoral Agreement for the Development of the Hydrogen Economy of 2020 (Ministry of Climate and Environment, 2020)—includes multiple goals concerning the development of hydrogen research (e.g. one of the key actions is 'Launching the domestic production of new hydrogen components, technologies based on the national scientific and research potential' [p. 12]). On the other hand, however, it identifies serious barriers for cooperation with HEIs, mainly in the commercialization of the results ('The parties recognise that there are barriers to translating research results into use of hydrogen technologies in practice in the form of commercialisation' [p. 14]). The document also bears the signature of representatives of more than 130 organizations, most of which represent business and state-owned companies; the list includes only five universities (four technical and one comprehensive). It illustrates the narrow focus of technical universities on STEM-related fields (Antonowicz, 2020) and comprehensive universities being focused primarily on abstract ideas and social principles. This leads to a lack of a holistic perspective in tackling the grand challenges.

Carrots—Economic Initiatives

In the absence of hard instruments such as governmental regulations designed specifically for the HE sector, numerous economic instruments (carrots) have been primarily designed to boost green policy. As a full member of the EU, Poland has access to considerable research funding through EU initiatives such as the European Green Deal (EC, 2019a). It encompasses a wide range of cross-sectoral initiatives linking strategies for the bio-economy or 'Europe for the Digital Age', as well as the search for synergies between programmes and initiatives: Horizon Europe, Digital Europe, Research Fund for Coal and Steel, Catching-up Regions Initiative, EU Renewed Industrial Policy Strategy and S-3 Vanguard Initiative. The European Commission (EC) has also prescribed guidelines for the design of national priorities and strategic objectives in the 2021–27 budget perspective, as reflected in Annex D to the European Commission's Country Report Poland 2019 (EC, 2019b). None of those initiatives is solely dedicated to the HE sector. However, they do not exclude HEIs from being actively engaged in the development of green technologies but do not also encourage them to be specifically involved. Financial resources are key to those initiatives and are being distributed through one of the public agencies, namely the National Centre for Research and Development (Narodowe Centrum Badań i Rozwoju—NCBiR), that finances applied research. NCBiR is directly accountable to the Ministry and, unlike the NCN (to be discussed later), can be considered a government's arm's-length organization.

Financial support by NCBiR is intended to encourage businesses to invest public resources in research and development ventures. Besides programmes addressed to businesses (the main beneficiary of the funds), those aimed at strengthening cooperation between scientists and businesses and support for scientific entities in the introduction of modern technologies, products and services to the market occupy an important place in NCBiR funding. In 2020, NCBiR allocated PLN 5 billion (about 1 million EUR) to support 947 research and development projects, and in 2021, 1199 projects received funding. The institution's budget increased from PLN 3.1 billion in 2016 to PLN 7.8 billion in 2021 (from 1.5 million EUR to about 6.5 million EUR). Predominant among the beneficiaries of completed projects in 2020 were business representatives, totalling 229 enterprises (178 small and medium-sized enterprises; 51 large enterprises; and 15 HEIs [6 universities, 6 research institutes and 3 research institutes of the Polish Academy of Sciences]). We do not have similar data for 2021, but we can assume that the proportions of businesses and universities benefiting from NCBiR programmes should be similar. Research objectives related to the climate change agenda arise from the multi-annual EU framework programmes, which are translated into general guidelines for the funding programmes implemented by NCBiR. However, in 2021, the NCBiR introduced a dedicated programme, 'Innovation Sapper', targeting projects related to climate change; nine such projects were carried out in 2021. NCBiR has also launched programmes for innovative projects in wind power, hydrogen technology and water management. Despite the lack of a formal climate change policy specifically designed for the HE sector in Poland, a government agency like the NCBiR is acting to steer universities and academics in certain strategic directions.

Sermons—Information and Policy Narrative

Wicked problems such as climate change require a long-term policy and a horizontal approach that activates the commitment of a wide range of policy actors (Sjöö & Callerstig, 2023). Policies are often preceded by debates and competing policy narratives that set the ground for the development of policy instruments (Head, 2022). Our brief examination of policy narratives provides compounding evidence that the Ministry of Education (earlier the Ministry of Science and Higher Education) has not recognized climate change or any other UN sustainable development goal as focal points for domestic policy discussions. Neither the Ministry nor other policy actors have proposed a specific policy agenda that takes a longer and broader view of HE and its potential contribution to society regarding climate change. This is the outcome of an amalgam of factors, including (as elaborated earlier) the political unwillingness to acknowledge the seriousness of climate change. Another factor is path dependency arising from the residual effect of the Humboldtian tradition to minimize relations with the external environment and, in fact, to reduce it to relations with the institution of the state and its agencies (Neave, 2002). Poland is not the only European country where universities are inspired by the Humboldtian ideas, but Polish universities rank amongst the most resilient institutions to the societal changes brought by modernization (Kwiek, 2015; Dobbins, 2015; Klenk et al., 2022).

Domestic policy debates in the field of HE are almost entirely dominated by internal issues such as funding, research evaluation, quality of education, internationalization and institutional governance (Ostrowicka et al., 2020; Dziedziczak-Foltyn, 2017; Antonowicz et al., 2020). Little attention has been paid to HE as part of its broader environment and its service to society. And where papers are available, they are confined to knowledge transfers and innovation for local and national economic development (IBnGR, 2010; KRASP, 2010; MniSW, 2015). A good proxy of the relatively low political priority given to non-economic goals, especially climate change, is the fact that among the eight thematic National Science Congresses (NKN), none was exclusively focused on them (Antonowicz et al., 2020). By far, it was a set of big thematic events organized about the most strategic aspects and policy goals of Polish HE that gathered thousands of academics and policymakers. Only one of them discussed the external relations of HE but focused exclusively on knowledge transfer, commercialization of research outcomes and technological innovation. Formally, the Ministry was largely responsible for setting the agenda for NKN, but the key policy themes were determined in consultation with representatives of the academic community. It shows that public debates around HE and policy discussion in that matter were driven almost exclusively by internal (HE field) problems.

The Institutional Level

Polish HEIs are publicly funded organizations that enjoy considerable institutional autonomy (Dobbins, 2015; Donina et al., 2022; Donina & Jaworska, 2022) to determine their strategic agency (Benneworth et al., 2017). Thus, to map institutional commitment to tackle climate change, we decided to examine institutional strategy as the indicator of underlying institutional priorities. Since 2011, all public HEIs have been mandated to have an institutional strategy (Dziedziczak-Foltyn, 2011). An institutional strategy is an important document, though it has zero impact on the level of public funding given to the universities. However, it manifests long-term institutional goals to which a university, as a community of scholars, declares its commitment.

However, these institutional strategies do not resemble those in the corporate sector because of the institutional governance model that allocates considerable powers to the academic community, as discussed earlier. Such strategic documents are formally prepared by the Rectors, but the content is typically a product of collegial effort, with a strong contribution from the academic community. It must also be approved by the university senate, which consists of representatives of the academic community, with the majority being professors.

In most cases, institutional strategies outline the mission statement and also underpin institutional values with frequent historical references (Anielska & Antonowicz, 2023). Polish HEIs are primarily historical institutions (Dobbins, 2015). So, their strategic documents do not really resemble corporate strategies and, as demonstrated in previous studies, are hardly useful as a managerial tool (Piotrowska-Piątek, 2015, 2018). However, they are significant carriers of values fundamental to particular academic communities. In short, if universities, as organizations, are committed to sustainable growth and particularly to tackling climate change, an institutional strategy would be the most appropriate place to declare it.

As discussed in the methodological section, we considered a sample of the top 20 public universities, as they are most prone to be engaged in addressing broader social issues. Initially, we identified references made to any of the selected grand challenges; in the second step, we established how important and specific these references are based on broader textual contexts. Last but not least, we tried to interpret these identified passages. Table 6.1 shows the results of the quantitative analysis performed concerning the number of references to the specified terms in each document. A reference to the carbon and ecological footprint was also coded as a mention of climate. Mentions of the demographic decline (declining student enrolment) were excluded from the 'ageing population' category. In most of the examined strategies, there were references to climate, sustainability and social responsibility. Social responsibility appeared in more strategies than the other categories, showing that this term is more recognizable to university decision-makers.

Table 6.1 shows that references to climate change were identified in 12 out of 20 strategies of universities. However, despite the seemingly strong presence of climate change in the institutional strategies, we assume that the role of those references is mostly ceremonial and suggests rather empty slogans rather than real engagement. References to grand challenges were expressed either generally ('major social and environmental challenges', [PG, 2020, p. 4]) or more specifically as a starting point for more general strategy assumptions ('uncertainty, geopolitical risks, difficult

Table 6.1 References to grand challenges in the universities' strategy documents

Case university (abbreviated)	*Climate change*	*Sustainable growth*	*Ageing*	*Social inequalities*	*Social responsibility*
AGH					
GUM					
PG					
PŁ					
PŚ					
PW					
PWr					
UAM					
UG					
UJ					
UŁ					
UMB					
UMK					
UMŁ					
UMP					
UP					
UPWr					
UŚ					
UW					
Uwr					
Total number of references to climate change issues	12/20	12/20	8/20	3/20	15/20

Source: Own study

macroeconomic outlook' [AGH, 2022, p. 11]). Regarding climate change, for example, the AGH strategy had a section on 'multi-faceted, sustainable development', in which the only mention of climate change referred to minimizing the carbon footprint of the university's management processes (p. 46). The same was true of the UŁ university strategy, in which climate change was mentioned only in the context of infrastructure adaptation (UŁ, 2021, p. 27). And in UMB's strategy, the only reference to climate change was related to the citation of the European Green Deal once at the very beginning of the document (UMB, 2021, p. 1).

It must be underlined that issues of climate change (and other grand challenges) were recognized and acknowledged in the institutional strategies, but they were not focal points for universities as such. Such findings stem from more contextual analysis in which 'climate change' appears rather parochial in comparison to apparently more prominent concepts such as academic excellence and entrepreneurialism. For example, in the PWr strategy, reference to climate change appeared twice (once in the context of Priority Research Areas and once in strategic goals related to the university's investment policy), while an entire section on 'Research for an Innovative Economy' was devoted to the commercialization of knowledge produced by the university (PWr, 2023, p. 14–15). In the UŚ strategy, 'Changes and environment and climate' was one of the headings of the Priority Research Area (UŚ, 2019, p. 13), but no changes were specified in any way. In the same strategy, on the other hand, the commercialization of research was conceptualized at the level of strategic goals and specific activities. It is exemplified by the designation of one of the goals as 'Cooperation with the socio-economic environment' (p. 32). However, there were some exceptions. PW, PG, Adam Mickiewicz University (UAM), Jagiellonian University (UJ), and Warsaw University (UW) approved strategies with the most extensive narrative regarding sustainable development. Moreover, these HEI strategies linked the climate change agenda to specific task domains: education (e.g. UAM); research (e.g. PW); infrastructure and investments (PWr, UAM, UJ, UW); and governance (e.g. establishing the Climate Council or [UJ] or the Climate Agenda [UW]). Interestingly, these universities also stand out from others regarding their performance and reputation; UJ and UW are ranked in the fifth hundred in the Academic Ranking of World Universities (ARWU) ranking. In other cases, we found references to grand challenges to be recognized as 'external circumstances' that shall be considered while planning institutional development but not as challenges that require strategic focus. So, they were located in the introduction to the strategies which outlined major social and environmental challenges ('uncertainty, geopolitical risks, difficult macroeconomic outlook', AGH, 2022, p. 11). It is also worth considering the strategies of UAM that were not only more specific than others but also envisaged infusing issues of climate change and growing social inequalities into teaching curricula (objective 2.1.3., p. 16). Some universities introduced separate all-university courses on sustainable development (UJ, p. 11; UŁ, p. 24; UW, p. 32), providing a detailed and more systematic study on this matter. Last but not least, UAM and UŁ promised to take up a more influential role in shaping public discourse, promoting knowledge and building public awareness about suitable development, which (from the context) could be read as symbolizing grand challenges.

The Individual Level

The most challenging aspect of the analysis was to map the individual activity of academics in tackling climate change. We believe the best available proxy for mapping the academic interest is a list of projects awarded funding by the NCN. The NCN is an autonomous public agency with a rather modest annual budget (300 million EUR) considering the size of the country; nonetheless, it remains the biggest single source of basic research funding in Poland. The NCN supports individual and team research projects, most of which are curiosity-driven studies submitted by individual academics. If anything provides a good view of the spectrum of the investigated research problems, it is the list of projects funded by NCN. However, our approach also has some reasoning limitations because we could identify only projects that qualified for funding. It is a tiny fraction of those submitted to the NCN, as the success rate ranges from 7% to 15%, depending on the call and panel. The analysis covered 341 successful applications during 2016–2022. Further, most academics conduct their research by either using other research funds (though the options are limited) or taking advantage of institutional resources, though these resources are usually very scarce. Nevertheless, we assumed that the NCN is a reliable proxy for assessing the academics' interest in grand challenges.

Overall, the NCN registered a fast upswing in interest in research topics related to climate change. In 2016, it had financed only three projects specifically focusing on climate change, but six years later, it was already 100. This shows scholars' rising interest in climate change. The rationale behind such popularity of studying issues related to climate change can vary significantly between individual projects. Nevertheless, in 2022, almost one-third of the funded projects were related to climate change. No other research topic had managed to capture such large attention, with a budget of around 300 million PLN (75 million Euros) or 25% of the NCN's total budget (Table 6.2).

All these projects between 2016 and 2022 were carried out by 270 universities and only 2 of them were non-public. The funded projects (Table 6.3) were awarded in the disciplinary fields of the natural sciences (36%) and engineering and technology (35%). The social sciences field also received a big share of projects focused on climate change (23%), demonstrating that it can be investigated independently from various angles.

Table 6.2 Research projects related to climate change funded by the NCN

Research project related to climate change funded by the NCN (N = 341)	*Number of total funded projects*	*Share of total funded projects*
2022	100	29.3%
2021	78	22.9%
2020	52	15.2%
2019	53	15.5%
2018	44	12.9%
2017	11	3.2%
2016	3	0.9%

Source: Compiled by the authors using data from the NCN 2016–2022

Table 6.3 Projects funded by the NCN by field of science

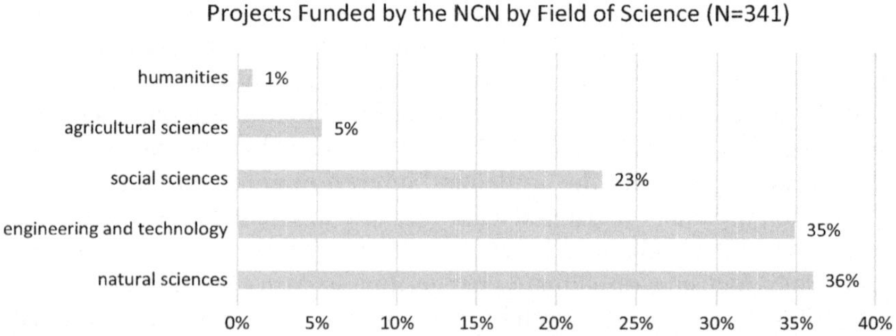

Source: Compiled by the authors using data from the NCN 2016–2022

Based on the analysis, we cannot conclude whether the conduct of research on climate change is linked to the scientific strategies of these universities or the result of the interest of individual scientists. However, since most universities generally include the climate change agenda in their institutional strategies without specifying what it stands for, we lean towards the second explanation. Undoubtedly, there is a wide range of individuals from many institutions who conduct studies on climate change, driven by curiosity and concern, and Polish universities provide them ample opportunity to do so.

Discussion and Conclusion

Tackling climate change has only very recently become a central topic in the Polish HE sector. The empirical analysis conducted in this study demonstrates that this is a rather bottom-up process inspired by the interests of individual academics, with little institutional engagement and rather passive policy measures.

The analysis shows that on the policy level, the Polish government uses only economic incentives (carrots). In practice, it simply follows the European principles on Green Policy, mostly by streaming (European) funding to facilitate research and innovation. However, two disclaimers have to be made in this context. The economic initiative is dedicated to facilitating green transformation (transition) in the private sector, which requires some form of partnership with universities. Still, considering the heavily underfunded Polish HE sector, these are substantial resources. If the public policy is whatever the government does or decides not to do, Polish governments have decided not to be overtly committed to tackling climate change. This is an ideological choice. However, they have done what is necessary to receive European funds—nothing more than that. The most compelling evidence is provided by the Ministry for Higher Education, which has not developed any specific policy agenda to address grand challenges in the field of HE. The government follows the European Green Policy only because this policy offers access to financial resources meant for the underfunded HE sector, though the universities receive only a small fraction of this funding. Besides, it neither attempts to introduce regulations (no sticks) that can impact public universities nor encourages them (no sermons) to be more active or stimulates them to be more mindful of climate issues.

The intensive policy debates about the agenda of HE reforms (2016–2017) suggested priority has been given to internal problems such as the role of humanities and social sciences, institutional funding and internal and external governance, as well as highly specific criteria research evaluation (Antonowicz et al., 2023). So, the Polish HE sector is defined as a largely hermetic system heavily influenced by a traditional academic oligarchy (Clark, 1983), in which internal actors mostly hold conversations with governments and among themselves. It largely contrasts with other systems of HE in Europe, which are more open to the influence of external stakeholders. This was perhaps an effective arrangement in the nineteenth century during the development of modern nation-states but has largely failed to live up to growing social expectations in current times. Other possible expectations include low public interest in climate change, which impacts policy narratives underpinning agenda setting. Instead of universities and academics adopting a more radical stance relating to grand challenges, as opinion leaders in the public discourse (as in other countries), there is no domestic tradition and little interest in thinking about HE in a broader environment and specifically in more instrumental terms. The economic engagement (still insignificant) has been partly forced upon universities by the recent policy measures and partly because this is a good source of income for them. Tackling climate change does not seem to tick any of those boxes.

This negative trend can be taken into the broader context of climate politics in Poland. From this perspective, HEIs are basically a reflection of a broader institutional and regulatory field in which tackling climate change is not a priority goal. According to the Climate Action Network Europe report, Centeran and Eaastern Europe (CEE) countries, and Poland among them, have the least ambitious climate policy goals of all EU countries (CANE, 2018). The Climate Change Performance Index (CCPI) ranks Poland 55th out of 67 countries analysed, mainly due to the lack of a long-term strategy to reduce Greenhouse Gas (GHG) emissions and an outdated energy mix based largely on fossil fuels (CCPI, 2024). Given the bigger picture and the lack of ambitious pro-climate policies in domains other than HE, it is hardly surprising that tackling climate change is not an important goal in either science-related public policy or institutional strategies.

This, however, does not mean that the academic community shows little interest or concern about climate change. On the contrary, the number of research projects funded by the NCN and also their share among all funded projects have been growing. Most of the calls of the NCN are open to any basic research project, reflecting individual curiosity, interest and concerns. Of course, the applicants choose research topics strategically to maximize their chances of funding, but it is also academics who evaluate the proposals. One way or the other, academics who are most active and successful research-wise find climate change a critical problem that requires in-depth studies.

We read the apparent inconsistency of policy inertia (Bonatti & Fracasso, 2019) and the spectacular growth in academics' interest in studying climate change which is occurring due to several factors. First is the disclosed political unwillingness of the Polish government to acknowledge and address climate change. The second is the historically inspired resistance of universities to be utilitarian to society for any purpose, even one as paramount as addressing climate change. However, this situation can easily change under the influence of the European Higher Education Area (EHEA) and particularly new structural developments like the European Universities Initiative (EUI) (Maassen et al., 2023) that put great emphasis on institutional relations with the external environment. Polish universities have joined the European Alliances enthusiastically and in a considerable number (Antonowicz et al., 2023), and with that comes a recalibration of the societal role. This includes greater responsibility of universities as socially engaged actors at all levels. The transnational character of the EUI draws closer attention to challenges that extend beyond national borders. Of course, it is a long process before the chronically underfunded and increasingly politicized Polish HE system moves from taking care of itself to addressing great social and economic challenges.

Historically, during the communist era, Polish universities were disconnected from broader society and its challenges; the democratic transformation (1990s) did not change this. It is how the 'republic of scholars' manifests their privileged status and exceptional role in society. As a seminal Polish philosopher picturesquely captured it, 'university is to do nothing useful' (Szahaj, 2018). Such account of university stays largely at odds with other European countries; Polish universities are not being transformed into organizational actors (Krücken & Meier, 2006), and neither do policymakers and academics (including institutional leaders) see them as entities with an endowed social agenda (Ramirez, 2010). Instead, the role of HEIs is to facilitate the work of individual academics in conducting their research and educating students as outlined by the Humboldtian model (Nybom, 2003), which they indeed do. Nevertheless, the skyrocketing interest in studying climate change shows that even if universities as organizations are rather reluctant to engage in this issue institutionally, they still provide a stable environment for academics to study problems they find particularly important, like climate change.

This raises the question: Is that enough from the HE sector to tackle climate change? It is probably not, considering the scale and seriousness of the problem and its massive impact on society, but some would say it is the best universities can do in the current circumstances. Tackling the grand challenges during the climate crisis means committing to grand modernization projects, including those related to energy transitions and the gradual decarbonization of the economy and society. Polish universities operate mainly in the scope of the prestige game, which makes their activities oriented within the academic field rather than towards the quest for pursuing external strategic goals. Consequently, it is hardly surprising that their participation in nationwide projects is negligible. Instead, the quest for 'world-class' status is some form of 'grand challenge' for both HE policy and universities, appearing to be far more important than other goals in the field of HE. The Excellence Initiative (IDUB) and the pouring of additional financial resources, together with a strong policy narrative, indicate that universities are more engaged in the race for global prestige rather than in the quest to address global challenges.

In recent years in Poland, climate policy has become a highly polarizing topic among major political actors (Biedenkopf, 2021), and the direction of regulation, including in the HE area, has been given by a right-wing government sceptical of European climate policies. We should also mention here the populist discourses that have dominated in recent years in Poland, especially in the right-wing media (Żuk & Szulecki, 2020), which, due to their anti-elitism or the perception of the EU as a threat to national sovereignty, have fostered scepticism about climate policies, which could spill over into various spheres, including HE. The new Polish government brings some hope for revising the public policy agenda and assigning higher priority to problems of paramount importance, following the green agenda of the EU. However, it takes time for the universities to overcome their historical hesitation and become socially engaged institutional actors. The biggest hope, however, comes with the surging research interest of individual academics in climate change. Regardless of their motivation, the issue is gaining more prominence through empirical findings, knowledge production and dissemination, both in closed academic circles but mostly among the general public. The critical mass of researchers and their alarming findings will soon become a turning point in a broader discussion about the role of the HE sector in tackling climate change. We are sanguine that considering increasing studies conducted in Polish universities, universities will gradually become more active and institutionally engaged in tackling climate change, which will eventually impact public policies too.

References

AGH. (2022). *The strategy of the AGH University of Science and Technology*. Retrieved December 17, 2023, from https://bip.malopolska.pl/agh,m,401336,strategia.html

Anielska, A., & Antonowicz, D. (2023). Strategies of higher education institutions towards postgraduate education in Poland. *Research in Post-Compulsory Education, 28*(4), 595–613.

Antonowicz, D. (2020). Understanding the development of technical universities in Poland. In L. Geschwind, A. Broström, & K. Larsen (Eds.), *Technical universities* (pp. 61–77). Springer.

Antonowicz, D., Budzanowska, A., Domalewski, J., Jaworska, M., Pokorska, A., & Wasielewski, K. (2023). Functioning and change of Polish universities within European universities initiative. In P. Poszytek & A. Budzanowska (Eds.), *European universities in Poland. Implementation of development strategy* (pp. 15–30). FRSE.

Antonowicz, D., Kulczycki, E., & Budzanowska, A. (2020). Breaking the deadlock of mistrust? A participative model of the structural reforms in higher education in Poland. *Higher Education Quarterly, 74*(4), 391–409. https://doi.org/10.1111/hequ.12254

Beerkens, E. (2008). The emergence and institutionalisation of the European higher education and research area. *European Journal of Education, 43*(4), 407–425.

Bell, D. (1976). *The coming of post-industrial society: A venture in social forecasting.* Basic Books.

Benneworth, P., Pinheiro, R., & Karlsen, J. (2017). Strategic agency and institutional change: Investigating the role of universities in Regional Innovation Systems (RISs). *Regional Studies, 51*(2), 235–248. https://doi.org/10.1080/00343404.2016.1215599

Białecki, I., & Dąbrowa-Szefler, M. (Eds.). (1994). *Changes in higher education in central European countries.* IFiS PAN.

Biedenkopf, K. (2021). Polish climate policy narratives: Uniqueness, alternative pathways, and nascent polarisation. *Politics and Governance, 9*(3), 319–400.

Bonatti, L., & Fracasso, A. (2019). Policy inertia, self-defeating expectations and structural reforms: Can policy modeling cope? *Journal of Policy Modeling, 41*(5), 943–962.

Brankovic, J. (2022). Why rankings appear natural (but aren't). *Business & Society, 61*(4), 801–806.

Brubacher, J. S. (1967). The autonomy of the university. *The Journal of Higher Education, 38*(5), 237–249.

Brzeziński, J., & Nowak, i. L. (red.). (1997). The idea of the university. *Poznań Studies in the Philosophy of the Sciences and the Humanities* (p. 50). Rodopi.

CANE (Climate Action Network Europe). (2018). *Off target. Ranking of EU countries' ambition and progress in fighting climate change.* CANE.

CCPI (Climate Change Performance Index). (2024). Poland. https://ccpi.org/country/pol/

Capano, G., & Howlett, M. (2020). The knowns and unknowns of policy instrument analysis: Policy tools and the current research agenda on policy mixes. *SAGE Open, 10*(1), 1–13.

Clark, B. (1983). *The higher education system: Academic organization in cross-national perspective*. University of California Press.

Cloete, N., Maassen, P., Fehnel, R., Moja, T., Gibbon, T., & Perold, H. (Eds.). (2006). *Transformation in higher education: Global pressures and local realities*. Springer Netherlands.

Cohen, M., James, M., & Johan, O. (1972). A garbage can model of organizational choice. *Administrative Science Quarterly, 17*(1), 1–25.

Cremonini, L., Westerheijden, D. F., Benneworth, P., & Dauncey, H. (2014). In the shadow of celebrity? World-class university policies and public value in higher education. *Higher Education Policy, 27*(3), 341–361. https://doi.org/10.1057/hep.2013.33

de Boer, H. F., Enders, J., & Leisyte, L. (2007). Public sector reform in Dutch higher education: The organizational transformation of the university. *Public Administration, 85*(1), 27–46.

de Boer, H., & van Vught, F. (2018/2020). Policy instruments in higher education. In P. N. Teixeira & J. C. Shin (Eds.), *The international encyclopedia of higher education systems and institutions* (pp. 2259–2265). Springer, Dordrecht.

Dennig, F., Budolfson, M. B., Fleurbaey, M., Siebert, A., & Socolow, R. H. (2015). Inequality, climate impacts on the future poor, and carbon prices. *Proceedings of the National Academy of Sciences, 112*(52), 15827–15832. https://doi.org/10.1073/pnas.1513967112

Dercon, S., & Christiaensen, L. (2011). Consumption risk, technology adoption and poverty traps: Evidence from Ethiopia. *Journal of Development Economics, 96*(2), 159–173. https://doi.org/10.1016/j.jdeveco.2010.08.003

DiMaggio, P. J., & Powell, W. W. (1983). The Iron cage revisited: Institutional isomorphism and collective rationality in organizational fields. *American Sociological Review, 48*(2), 147–160. https://doi.org/10.2307/2095101

Dobbins, M. (2015). Exploring the governance of polish public higher education: Balancing restored historical legacies with Europeanization and market pressures. *European Journal of Higher Education, 5*(1), 18–33. https://doi.org/10.1080/21568235.2014.969289

Donina, D., & Jaworska, M. (2022). Higher education governance in Poland: Reform pathway from the communist regime to law 2.0. *Higher Education Policy, 37*(1), 40–58. https://doi.org/10.1057/s41307-022-00293-x

Donina, D., Pokorska, A., Antonowicz, D., & Jaworska, M. (2022). A cultural perspective of higher education governance reform in Poland: Divergent interpretations by rectors across distinct categories of universities. *Journal of Higher Education Policy and Management, 44*(6), 596–612. https://doi.org/10.1080/1360080X.2022.2103940

Dye, T. (1972). *Understanding public policy.* Prentice-Hall.

Dziedziczak-Foltyn, A. (2011). Imperatyw rozwoju a kondycja myślenia strategicznego o polskim szkolnictwie wyższym (i nauce) w dobie transformacji systemowej. *Nauka i Szkolnictwo Wyższe, 2*(38), 88–106.

Dziedziczak-Foltyn, A. (2017). Reforma Szkolnictwa Wyższego w Polsce w Debacie Publicznej. In *Bilans Dyskusji o Uniwersytetach (1990–2015)*. UŁ.

Enders, J. (2014). The academic arms race. In A. Pettigrew, E. Cornuel, & U. Hommel (Eds.), *The institutional development of business schools* (pp. 155–175). Oxford University Press.

European Commission. (2019a). Communication from the commission to the European parliament, the European Council, the Council, the European Economic and Social Committee and the Committee of the Regions- The European Green Deal. Retrieved December 17, 2023, from https://eur-lex.europa.eu/legal-content/EN/TXT/?uri=COM%3A2019%3A640%3AFIN

European Commission. (2019b). Communication from the commission to the European parliament, The European Council, The Council, The European Central Bank and the Eurogroup. Retrieved December 17, 2023, from https://commission.europa.eu/system/files/2019-02/2019-european-semester-country-report-poland_en.pdf

Gornitzka, Å., Maassen, P., & De Boer, H. (2017). Change in university governance structures in continental Europe. *Higher Education Quarterly, 71*(3), 274–289.

Grzymala-Busse, A. (2018). Poland's path to illiberalism. *Current History, 117*(797), 96–101.

GUS. (2023). *Szkolnictwo wyższe i ich finanse.* GUS.

Hazelkorn, E. (2011). *Rankings and the reshaping of higher education.* Palgrave Macmillan.

Hazelkorn, E. (2014). Rankings and the global reputation race. *New Directions for Higher Education, 168*, 13–26. https://doi.org/10.1002/he.20110

Head, B. W. (2022). Debates in public policy – Problem framing, knowledge and interests. In B. W. Head (Ed.), *Wicked problems in public policy* (pp. 7–19). Palgrave Macmillan.

Huisman, J. (2009a). Coming to terms with governance in higher education. In *International perspectives on the governance of higher education* (pp. 27–36). Routledge.

Huisman, J. (2009b). *International perspectives on the governance of higher education.* Routledge. https://doi.org/10.4324/9780203883358

IBnGR. (2010). *Strategia rozwoju szkolnictwa wyższego w Polsce do 2020 roku.* IBnGR & EY.

Kaldeway, D. (2018). The grand challenges discourse: Transforming identity work in science and science policy. *Minerva, 56*(2), 161–182.

Klenk, T., Antonowicz, D., Geschwind, L., Pinheiro, R., & Pokorska, A. (2022). Taking women on boards: A comparative analysis of public policies in higher education. *Policy Reviews in Higher Education, 6*(2), 128–152.

KRASP. (2010). *Strategia rozwoju szkolnictwa wyższego 2010–2020*. FRSP.

Krücken, G., & Meier, F. (2006). Turning the University into an organisational actor. In G. S. Drori, J. W. Meyer, & H. Hwang (Eds.), *Globalization and organization: World society and organizational change* (pp. 241–257). Oxford University Press.

Kwiek, M. (2015). The unfading power of collegiality? University governance in Poland in a European comparative and quantitative perspective. *International Journal of Educational Development, 43*(C), 77–89. https://doi.org/10.1016/j.ijedudev.2015.05.002

Maassen, P., & Olsen, J. P. (Eds.). (2007). *University dynamics and European integration*. Springer.

Maassen, P., Stensaker, B., & Rosso, A. (2023). The European university alliances – An examination of organizational potentials and perils. *Higher Education, 86*(4), 953–968. https://doi.org/10.1007/s10734-022-00951-4

Marginson, S. (2010). Global comparisons and the university knowledge economy. In L. M. Portnoi, V. D. Rust, & S. S. Bagley (Eds.), *Higher education, policy, and the global competition phenomenon. International and development education* (pp. 29–41). Palgrave Macmillan.

MEN. (2022). Polityka Naukowa Państwa [National Science Policy]. Warszawa 2022. Retrieved December 17, 2023, from https://www.gov.pl/attachment/c47d37f9-eaab-4701-adce-84cf7585013e

Meng-Hsuan, C., Jungblut, J., Ravinet, P., & Vukasovic, M. (2017). Higher education governance and policy: An introduction to multi-issue, multi-level and multi-actor dynamics. *Policy and Society, 36*(1), 1–15.

Ministry of Climate and Environment. (2020). Porozumienie sektorowe na rzecz rozwoju gospodarki wodorowej w Polsce [Sectoral Agreement for the Development of the Hydrogen Economy]. Retrieved December 17, 2023, from https://www.gov.pl/attachment/28df7f48-feac-4eb0-9b47-308c0aa6e597

Ministry of Climate and Environment. (2021). Polityka Energetyczna Polski do roku 2040 [Polish Energy Policy 2040]. Warszawa. Retrieved December 17, 2023, from https://isap.sejm.gov.pl/isap.nsf/download.xsp/WMP20210000264/O/M20210264.pdf

Ministry of Economic Development and Technology. (2021). Polityka Przemysłowa Polski [Poland's Industrial Policy]. Warszawa. Retrieved December 17, 2023, from https://www.gov.pl/attachment/555c07a8-4d95-49af-9ec1-282fafdbcac5

MNiSW. (2015). Program Rozwoju Szkolnictwa Wyższego na lata 2015-2030. Warszawa. Retrieved December 17, 2023, from https://dn.po.edu.pl/index.php/program-rozwoju-szkolnictwa-wyzszego-i-nauki-na-lata-2015-2030

Neave, G. (2002). The stakeholder perspective historically explored: Higher education in a globalising world. In J. Enders & O. Fulton (Eds.), *International trends and mutual observations a festschrift in honour of Ulrich Teichler* (pp. 17–37). Springer.

Nybom, T. (2003). The Humboldt legacy: Reflections on the past, present, and future of the European university. *Higher Education Policy, 16*(2), 141–159.

Ostrowicka, H., Spychalska-Stasiak, J., & Stankiewicz, Ł. (2020). *The dispositif of the university reform. The higher education policy discourse in Poland.* Routledge.

PG. (2020). Gdańsk University of Technology Strategic Framework 2020 – 2030. Retrieved December 17, 2023, from https://pg.edu.pl/files/2021-03/Strategia_PG_2020-2030.pdf

Piotrowska-Piątek, A. (2015). Analiza strategii rozwoju szkół wyższych w Polsce. *Organizacja i Zarządzanie: kwartalnik naukowy, 2*(30), 103–116.

Piotrowska-Piątek, A. (2018). *Współpraca z otoczeniem w misjach i strategiach szkół wyższych.* CeDeWu.

Polanyi, M. (1962). *Personal knowledge: Towards a post-critical philosophy.* Chicago University Press.

PWr. (2023). Strategy of Wrocław University of science and technology 2023–2030. Retrieved December 17, 2023, from https://pwr.edu.pl/fcp/LGBUKOQtTKlQhbx08SlkTUhZeUTgtCgg9ACFDC0RGS3xSFVZpCFghUHcKVigEQUw/1/public/2023/docs/strategia__pwr_2023_22-06-23.pdf

Ramirez, F. O. (2010). Accounting for excellence: Transforming universities into organizational actors. In L. M. Portnoi, V. D. Rust, & S. S. Bagley (Eds.), *Higher education, policy, and the global competition phenomenon, international and development education* (pp. 43–58). Palgrave Macmillan.

Ramirez, F. O., & Tiplic, D. (2014). In pursuit of excellence? Discursive patterns in European higher education research. *Higher Education, 67*(4), 439–455. https://doi.org/10.1007/s10734-013-9681-1

Shaw, M. A. (2019). Public accountability versus academic independence: Tensions of public higher education governance in Poland. *Studies in Higher Education, 44*(12), 2235–2248. https://doi.org/10.1080/03075079.2018.1483910

Sjöö, K., & Callerstig, A. C. (2023). The challenges in integrating horizontal perspectives in sectoral policy evaluation. *Policy Studies, 44*(2), 174–194.

Szahaj A. (2018). *Uniwersytet jest po nic.* Retrieved December 17, 2023, from https://www.rp.pl/opinie-polityczno-spoleczne/art1864731-szahaj-uniwersytet-jest-po-nic

Thornton, P. H., & Ocasio, W. C. (2008). Institutional logics. In R. Greenwood, C. Oliver, K. Sahlin, & R. Suddaby (Eds.), *The SAGE handbook of organizational institutionalism* (1st ed., pp. 99–129).

Trnavcevic, A. (2010). Introduction to section A: The role of higher education in a post-conflict society. *Interchange, 41*(2), 99–102.

Trowler, P. (2002). Higher education policy and institutional change: Intentions and outcomes in turbulent environments. *Higher Education, 47*(2), 253–255. https://doi.org/10.1023/B:HIGH.0000016667.05912.f4

UŁ. (2021). The strategy of the University of Lodz 2021–2030. Retrieved December 17, 2023, from https://www.uni.lodz.pl/o-uniwersytecie/misja-wizja-wartosci-strategia

UMB. (2021). *The development strategy of the Medical University of Bialystok for the years 2021–2030*. Retrieved December 17, 2023, from https://www.umb.edu.pl/photo/pliki/administracja/dzial_nowoczesnych_metod_i_technik_ksztalcenia/dd/strategia/22.2021_strategia_umb_na_lata_2021-2030-22.pdf

UŚ. (2019). Development strategy of the University of Silesia in Katowice for 2020–2025. Retrieved December 17, 2023, from https://us.edu.pl/strategia-rozwoju-uniwersytetu-slaskiego-w-katowicach-na-lata-2020-2025/

Vedung, E. (1998). Policy instruments: Typologies and theories. In M. Bemelmans-Videc, R. C. Rist, & E. Vedung (Eds.), *Carrots, sticks, and sermons: Policy instruments and their evaluation* (pp. 21–58). Transaction Publishers.

Wiejski, P. (2022). *The accounting trick behind Poland meeting its EU renewable energy target*. Retrieved December 17, 2023, from https://notesfrompoland.com/2022/01/25/the-accounting-trick-behind-poland-meeting-its-eu-renewable-energy-target/

Yudkevich, M., Philip, G. A., & Salmi, J. (2023). *Academic star wars: Excellence initiatives in global perspective*. MIT Press.

Żuk, P., & Szulecki, K. (2020). Unpacking the right-populist threat to climate action: Poland's pro-governmental media on energy transition and climate change. *Energy Research & Social Sciences, 66*, 101485.

CHAPTER 7

Green-Dealing in Czech Higher Education: Agenda Only for the Initiated?

Jan Kohoutek (1978–2024)

Introduction

In recent decades, rapid changes in the earth's climate, involving an overall increase in temperatures and other environmental effects, have become a reality in the lives of all. Driven by carbon dioxide and other greenhouse gas emissions substantially produced by human activity, climate change has emerged as one of the world's most serious concerns. Its acuity is due to its actual effects on the world's population, manifested in rising temperatures, sea levels, or mounting biodiversity losses, as it is difficult to solve. Climate change is complex and has multiple causes and disproportionate impacts. Further, interventions in one area may bring unexpected changes in others (McCowan, 2020). Moreover, such changes may cancel out net positive effects, having strong environmental, economic, political, and social consequences (Filho, 2010).

J. Kohoutek (Deceased) (✉)
Department of Public and Social Policy at the Faculty of Social Sciences, Charles University, Prague, Czech Republic
e-mail: romulo.m.pinheiro@uia.no

J. Kekäle, R. Pinheiro (eds.), *Higher Education Policy for Tackling Climate Change*, https://doi.org/10.1007/978-3-031-78479-8_7

As such, climate change qualifies as a super wicked problem (Levin et al., 2012). Drawing on the classic work of Rittel and Webber (1973), Levin et al. argue that "super wicked" represents a new class of global environmental problems because of four distinct qualities, namely, time for a solution is running out, those who cause the problem are also supposed to provide a solution, the central authority needed is weak or non-existent, and irrational discounting pushes responses into the future (2012, p. 124). In this respect, adaptation and mitigation have been designed as pivotal in addressing climate change (Alves et al., 2020). Mitigation addresses the causes of climate change by deploying (investments in) renewable energy, carbon trading, geo-engineering, and educating the citizenry. Adaptation objectives address the impacts of climate change through an adjustment in natural or human systems (agriculture, housing, etc.) to moderate harm or exploit opportunities and synergies (Alves et al., 2020; McCowan, 2020). In Europe, the Green Deal stands out as a prime contemporary initiative of the European Union (EU), involving adaptation and mitigation for combating climate change. Set as the blueprint for transformational change, the Green Deal obligates all 27 EU Member States to turn the EU into the first climate-neutral continent by 2050. To achieve this, the member states have pledged to reduce emissions by at least 55 percent by 2030, vis-a-vis 1990 levels (European Commission, 2019).

Universities are uniquely positioned to bring mitigation and adaptation processes to the fore and thus help toward climate change solutions (Filho et al., 2019, 2020; McCowan, 2019). This is due to their educational activities shaping human mindsets, university-based research producing key innovations in technologies and beyond, and universities' broader societal mission being geared toward raising awareness of the issue and promoting sustainable strategies or programs addressed to various target audiences. Despite universities' significant potential, their actual role in climate change issues seems more ambiguous. Universities indeed educate students about global warming and climate change, but the effects are long-term and nebulous; moreover, many country leaders, including but not limited to highly educated high-level politicians, adopt a reserved approach when making binding commitments. Second, many university departments still research technologies that enable the continued exploitation of resources and extension of fossil fuel production. Third, competing with each other for status, universities tend to expand in size and scope, which often means building new infrastructure and pushing material and energy requirements to new heights. Finally, universities contribute to fossil fuel emissions through the global travel of academic staff and students on exchange programs (McCowan, 2020; McCowan et al., 2021).

Hence, the overall picture is mixed, and currently, we lack a clear knowledge and understanding thereof. On the one hand, relevant policies are required, involving universities and other higher education institutions (HEIs) in positively addressing climate change issues. On the other hand, there is little insight into the extent to which such policies, proposing "green agendas," are part of higher education (HE) policymaking systemically or institutionally and the effects thereof (Filho et al., 2019).

In short, more insight is needed into how (or whether) green issues make it to the HE agenda, how much priority they are given, and with what effects. The lack of such insights poses the following question: *how are climate change and related environmental issues dealt with in agenda-setting in the higher education sector and, more widely, in public policymaking?* A related question is *what are the underlying drivers, measures, and instruments used for the green and digital transitions of the higher education sector?* These questions are addressed concerning the Czech Republic (CR) in the form of a case study. I argue that answering these questions will help in achieving the aim of the chapter, which is an exploration of the extent to which climate change issues figure as an agenda in HE policies in the CR. More broadly, the chapter attempts to set the resulting findings into a broader framework of climate protection (in)actions and draw relevant international implications.

Against this backdrop, this chapter continues by presenting its focus and methodology, followed by insights into agenda-setting instruments that conceptually underpin the study. Next, the chapter lays out fundamental facts and figures about Czech HE to contextualize the case. This is followed by the presentation of the rationale and contents of the Czech Recovery and Resilience Plan (RRP) in connection with the country's strategic documents, namely "the Strategic Framework CR 2030," "the Strategic Plan for Higher Education 2021+," and "National Priorities of Research.[1]" After framing the issue of climate mitigation and protection in strategic terms, the study proceeds to identify concrete measures and instruments of the RRP used for the green and digital transition of the HE sector. Examples of individual HEIs and their corresponding actions are provided. Next, the effects of the RRP and corresponding HEI actions are analyzed, followed by a broader discussion on the utility of the HE green

[1] The full title of this document is "National Priorities of Oriented Research, Experimental Development and Innovations."

transition in policy and political terms. Particularly, the tensions between grand investments, actual delivery, and mounting (domestic) EU criticism are highlighted. The chapter concludes by providing tentative pointers for national and international research into green transition in HE.

Putting the Czech Republic into Focus

The CR presents a fitting case for enquiring into climate change agendas in HE policies for both factual and empirical reasons. Although there is some literature, primarily economic, on the CR's green transition (Marek & Krejza, 2023; Slavík et al., 2018; Siuda, 2023; Sila, & Frohm, 2023), including green public procurements (Plaček et al., 2021; Páleníková et al., 2022), complemented by studies of benefits of education for sustainable development (Dlouhá & Dlouhý, 2014; Dlouhá & Pospíšilová, 2018), none of these works deal analytically with climate-saving agendas in Czech HE policy settings, including the role of universities. Second, empirical data on emission production from 2020 show that the CR's economy is still very carbon-intensive and generates 3.5 percent of the EU's total greenhouse (GHG) emissions (EPRS, 2021). With high levels of energy-intensive industry and persistent reliance on coal-intensive supplies, domestic emission intensity is significantly higher than the EU average, though it is on a downward trend (EPRS, 2021). In this regard, meeting the goal of the EU's Green Deal to reduce emissions by at least 55 percent by 2030 is a challenging task.

To accelerate emission reduction in line with the Green Deal, the Czech government has put in place in 2021 the RRP valued at EUR 7.036 billion (approximately CZK 180 billion)[2] (EC, 2021). The RRP comprises ambitious measures facilitating the Czech economy's digital transformation and green transition during 2021–2026. It contains six priority areas, broken down into 26 components. The six areas are digital transformation; physical infrastructure and green transition; ***education and labor market***; institutions, regulation and business support; ***research, development and innovation***; and health and resilience of the population (MPO, 2021). The RRP was updated in 2023 with the addition of CZK 50 billion, out of which CZK 17.4 billion is allotted to a new priority area called "REPowerEU" (MPO, 2023). HE is central to the priority areas of education with labor market

[2] To be drawn from the EU recovery programme. The Czech RRP also includes measures worth CZK 11.3 billion (EUR 0.44 billion) to be funded from national sources.

and research, development, and innovation. As will be shown, HE also plays an important role in the new REPowerEU area.

So far so good; however, the CR has been known to be a reluctant reformer when it comes to the achievement of EU development goals (Meyer-Sahling, 2011). Hence, it is worth studying to what extent climate change agendas, particularly in the form of the RRP, have been on the Czech policymaking agenda in HE and have so far escaped research scrutiny. As suggested in the introduction, this chapter thus aims to explore the extent to which climate change issues are set as an agenda in HE policies in CR, system-wise. Correspondingly, the aim is also to identify the agenda-setting instruments in use, analyze their effects, and draw wider, internationally relevant implications.

Methodology

Being exploratory, this chapter makes use of the review of relevant documents based on desk research. Desk research refers to "the process of collating and coding existing information for analysis, without direct contact between researchers and research participants" (Green & Cohen, 2021, p. 2) with sources including archival and government documents, NGO reports, articles, strategic documents of different kinds, annual reports, statistics, and assessments.

Correspondingly, the desk research for this study comprised the review of a wide range of relevant sources in English or Czech language. They were research articles and books on the theme, along with literature on agenda-setting. Other sources comprised the Czech government and Ministerial strategic policy plans, the RRP's website, websites of selected Czech HEIs, the EU ex-ante assessment of the RRP, statistical data, and sociological surveys. All the sources yielded vital pieces of information on green agendas in Czech HE, enabling the identification of the corresponding drivers, instruments, and effects.

Agenda-Setting: Conceptual Significance and Application

As an initial phase of the policymaking cycle, *agenda-setting* refers to the process by which some problems come to public attention at given times and places (Weiss, 1989, p. 118), involving the politics of selecting issues for active consideration (Cobb & Ross, 1997). Agenda-setting processes often occur within the purview of government activities (Howlett, 1997; Bali & Halpin, 2021) and encompass the governmental agenda (the problems that decision-makers pay serious attention to) and the decision agenda (problems up for an active decision) (Cairney, 2012, p. 183). This suggests that many policy problems can be taken up as an agenda, but the government considers only very few at a given time (becoming part of the governmental agenda). To control the number and nature of problems qualifying for the agenda (filter out problems), governments typically employ a range of tools. Policy tools/instruments are the subject of deliberation and application at every stage of the policy process, including agenda-setting (Shivakoti & Howlett, 2022). It is argued that by examining the toolsets available to the government at the agenda-setting stage, we can better assess and understand how governments control or attempt to control issue prominence while trying to manage or direct their own policy timetables and agendas (Shivakoti & Howlett, 2022).

A corresponding typology has been proposed to study the effects of toolsets (i.e., mixes of tools) during the agenda-setting phase (Bali & Halpin, 2021; Shivakoti & Howlett, 2022). Drawing on Hood's (1986) four-fold classification of the resources that governments use to effect change, that is, *nodality* (information), *authority*, *treasure*, and *organization* (NATO), the classification is recast to exemplify typical agenda tools for each resource type, complemented by aims to manage the corresponding government policy demands and preferences during the agenda-setting stage. The resulting typology of agenda-setting tools is shown in Table 7.1.

A scrutiny of the typology reveals distinct strategies by which governments aim to shape the agenda-setting phase, ranging from issue management, issue permission through issue preference (by fostering demands), and issue imposition. The typology, along with the corresponding insights, is used for the case analysis by decomposing the processes of adoption of the Green Deal as an agenda in Czech HE policymaking, including identification of the aims, instruments in use, and their effects.

Green Deal in Czech Higher Education Policy Agenda

Contextual Backdrop

Before analyzing the Green Deal as an agenda in Czech HE, key characteristics that underlie the Czech system of HE need to be presented. This

Table 7.1 Typology of agenda-setting instruments

Resource type	*Aim*	*Example of tool applied*
Nodality (information)	Managing the demands making it to the governmental agenda (routinize)	Establishing workgroups, taskforces, government communications
Organization	Permitting issues (back) on the governmental agenda at set intervals (regularize)	Scenario planning, strategic foresight
Treasure	Fostering policy demands to help shape governmental agenda proactively (generate)	Funding of research, procuring research
Authority	Unilaterally install government issue priority on the agenda in the absence of consultation (esp. after the crisis) (impose)	Periodic reviews, establishing mandates

Source: Adapted from Bali and Halpin (2021) and Shivakoti and Howlett (2022)

is to set the case in a wider international context. After the watershed year of 1989, academic freedoms were swiftly renewed at HEIs, and program curricula, as well as research, were freed of the Communist dogmas. The democratization of HEIs was followed by the introduction of formula funding for study places (1993), mandatory program accreditation (1999), entry of new privately run providers (1999), and the phasing in of the BA-MA-PhD degree structure along the Bologna template in the early to mid-2000s. The demand for study places was high (especially in attractive fields, such as law, economics, and medicine), often outstripping available institutional capacities. Between 1990 and 2010, the total number of students enrolled almost quadrupled (from 110,000 to 395,000), with universal access to HE achieved in the second half of the 2000s. Since then, the student numbers have been on the decline due to demographic change, reaching 311,000 in 2017 and decreasing further to 304,000 in 2022 (CSVŠ, 2018; ČSÚ, 2022).

Currently, the system registers 26 public HEIs, 30 private HEIs, and two state HEIs.[3] Most public HEIs offer accredited doctoral programs (PhD), giving them the official status of a university. Private HEIs are

[3] The state HEIs—The Police Academy and The University of Defence—are organizationally and financially subordinated to the Ministry of Interior and the Ministry of Defence, respectively.

small in capacity and often offer vocation-oriented Bachelor programs. Only private HEIs charge tuition fees, whilst state and public HEIs are financed from the state budget. The total subsidy for study places at public HEIs exceeded CZK 21 billion (EUR 0.84 billion) in 2017, representing 0.43 percent of the country's GDP (below the OECD average). The governance of Czech HEIs is predominantly collegial, with the university and faculty senates composed of academic and student representatives. At the system level, the Ministry of Education is legally obligated to consult on major issues in HE, including budgetary conditions, with two HE representative bodies, the Rectors' Conference, and the Council of HEIs (including the Students' Chamber).

Green Deal, Policy Strategies, Recovery, and Resilience Plan

Adopted in 2020, the European Green Deal has been heralded as "*Europe's man on the moon moment*," as it aims to reduce GHG emissions by 55 percent by 2030 on the way to reaching climate neutrality by 2050 (Hainsch et al., 2022). To aid in realization, this ambitious aim of the Green Deal is accompanied by the corresponding goals that include, inter alia, the supply of clean, affordable, and secure energy, industry modification to a circular economy, preservation and restoration of biodiversity and acceleration of the shift to sustainable and smart mobility (European Commission, 2019). The role of education is acknowledged as schools and HEIs "are well placed to engage with pupils, parents, and the wider community on the changes needed for a successful transition" (European Commission, 2019, p. 19). The engagement about climate change should proceed through information sharing and inspiration, holding real or virtual discussion forums, and capacity building to facilitate grassroots climate mitigation initiatives, including the exchange of good practices. To such ends, (HE) curricular content should be digitalized and (higher) education buildings renovated. While the building renovations are expected to result in less intensive raw material use and greater energy efficiency, the new digital infrastructure, including clouds, ultra-fast networks, and artificial intelligence solutions, should "facilitate evidence-based decisions and expand the capacity to understand and tackle environmental challenges" (European Commission, 2019, p. 18).

Representing the most extensive action of the EU in climate protection, the Green Deal is a complex initiative whose goals cut across a range of sectors, including HE. The latter has a specific position due to its capacity to educate about and research environment-friendly solutions. To implement the goals of the Green Deal, the Deal is accompanied by a stimulus package worth EUR 2.018 trillion at current prices. Besides the EU's budget of EUR 1.211 trillion for 2021 to 2027, the package also includes the NextGenerationEU recovery program worth EUR 806.9 billion. This is to boost EU member states' economies, struck by the COVID-19 pandemic recession, and steer their modernization toward greater resilience, digital transitions, and fighting climate change in tune with the Deal goals (European Commission, 2019).

The Green Deal presented itself as a strategic opportunity high on the policy agenda of the CR government as an EU Member State. Between 2020 and 2021, over EUR 7 billion was offered to the CR on the condition of putting forth a relevant modernization plan in line with the Green Deal, further subject to the Commission's approval. Given the EU's institutional significance, political clout, and the level of funding offered, the Green Deal turned into a policy driver, generating a response by the making of the modernization plan on the part of the Czech government and subordinate state administration units. The entire Green Deal agenda, including the making of an appropriate modernization plan for CR, came to be coordinated centrally by the Czech government with the management of issue areas to qualify for funding from the plan vested in the Czech line ministries. The agenda management thus took a top-down approach, and the years 2020 and 2021 were marked by intense activities of line/sectoral ministries to draw up contributions to the national modernization plan, justifying their sectoral financial share.

In strategic terms, the overarching document of the country, titled "Strategic Framework: Czech Republic 2030," provides for climate change mitigation and decarbonization of the economy, including reduction of GHG emissions and increasing digitalization and robotization. The framework also acknowledges the central role of research, development, and innovations for technological advancements, the importance of the circular economy, higher value-added industries, and low-carbon technologies. In short, the framework has many aspects of the country's green transition covered (including adaption of cities and housing).

However, concerning the HE sector, the green transition is covered very nebulously in "The Strategic Plan of the Ministry for Higher Education for the Period from 2021" (MEYS, 2020). This plan, compiled in 2020 and updated annually, represents the main strategic document for Czech HE and has six priority goals:

1. Developing competencies relevant for life in the twenty-first century,
2. Improving the availability and relevance of flexible forms of education,
3. Improving the efficiency and quality of doctoral studies,
4. Strengthening strategic management and the effective use of capacities in research and development at HEIs,
5. Building capacity for the strategic management of HE, and
6. Reducing the administrative burden on the staff of HEIs.

Thus, none of the plan's priorities refers to climate saving or green issues.[4]

On the other hand, climate change adaptation is referenced under the "Smart Investments" priority in the Innovation Strategy of the Czech Republic (2019–2030). Having a bearing on setting nation-wide research priorities, the strategy also explicitly addresses issues of digitalization of state services and administration. Finally, there are a few scant references to decarbonization, lowering emissions of GHG, ecological transport, and green infrastructure in "the National Priorities of Oriented Research, Experimental Development, and Innovations" (CR Government, 2012), which is the country's main research-specific strategic document valid till now.

Importantly, none of the strategies dealt with is binding legally or otherwise (assessment-wise), which may be why strategic policy documents are so popular in the CR (there are literally hundreds of them). Still, given the significance of the green agenda and the potential role of the Czech universities therein, not least due to the astronomic level of EU funding available, one would expect more than a few passing references in major strategies on HE; the annual updates of the HE Strategic Plan do not mention green transformation either.[5]

[4] Except for "the contribution of HEIs to the adaptation to climate change and ecological aspects of their operation, incl. reducing their carbon footprint," which is a two-liner listed last in the Plan among other significant themes in HE.

[5] Scant referencing to the green or digital transformation in strategic policy documents *central to higher education* might have been the reason why, for the sake of addressing the Green Deal agenda through the National Recovery Plan, the Department of Higher Education of the Ministry of Education was tasked with identifying such references in some other strategies, more distant to the HE field (as there are also some to choose from in CR).

The reaction of the Czech government to the EU's Green Deal stimulus package took the form of "the Recovery and Resilience Plan." The CR's RRP was submitted to the European Commission on June 1, 2021, securing the final approval, including funding allocation, on September 6, 2021. The originally approved RRP has six priority areas: digital transformation; physical infrastructure and green transition; education and labor market; institutions, regulation and business support; research, development and innovation; and health and resilience of population (EC, 2021). Each of the priority areas comprises several components (subareas), amounting to 26 components in the original RRP. The RRP funding is close to CZK 180 billion, corresponding to EUR 7.036 billion (also including measures worth CZK 11.3 billion (EUR 0.44 billion) from national sources).

The RRP envisages forward-looking measures facilitating the digital transformation (22 percent of the RRP budget) and green transition (42 percent of the RRP budget) of the Czech economy. The implementation of the RRP priorities should be finalized by 2026. The RRP was updated in autumn 2023, with the update providing an additional CZK 50 billion, out of which CZK 17.4 billion is allotted to a new priority area titled "REPowerEU." This new priority comprises reforms and adjustments in the energy sector, including affordable housing and the initiation of communal energy sharing. It is generally aimed at strengthening the country's energy security and expediting its energy transformation toward more sustainable approaches. REPowerEU has seven components, including the adaptation of Czech HEIs to green skills and sustainability. All the priority areas and components of the updated RRP can be found in the annex to the chapter (Table 7.2).

Recovery and Resilience Plan: Areas, Components, and Effects on Higher Education

As suggested, the RRP has six main areas, each sub-divided into components (subareas) under the administrative responsibility of a particular line ministry. The Ministry of Education administers the following four components also pertinent to the HE sector and public HEIs:

In 2021, the author of this chapter was among those externally commissioned to aid in this undertaking.

- 3.1 Innovation in education in the context of digitalization
- 3.2 Adaptation of the capacity and orientation of school programs
- 5.1 Excellent R&D in priority areas of public interest in healthcare
- 7.4 Adapting HEIs—promoting green skills and sustainability in HE.

Each component has its own goals, formulated as follows:

- 3.1. revision of the curriculum of primary and secondary schools, support of ICT education and sensible application of digital technologies, implementation of revised curricula including DigCompEdu framework, prevention of digital exclusion, and increasing the level of institutional digitalization.
- 3.2. transformation of HEIs to adapt to new forms of study and changing labor market needs, school support, support of catch-up classes, investments in selected university campuses (Albertov Campus of Charles University, Mephared 2 Campus of Charles University, and Biopharma Hub of Masaryk University).
- 5.1. support of research in specific biological, medical, and social fields, with a focus on infectious diseases, oncology, neurology, metabolic, and cardiovascular diseases, as well as on socio-economic impacts of systemic health risks.
- 7.4. transformation of HEIs to adapt to the changing circumstances in the labor market, creating strategies of sustainable and ecologic transformation, and making strategic partnerships.

Importantly, subareas 3.1, 3.2, 5.1, and 7.4, along with the corresponding goals, have been set for financial support in the form of targeted or block grant funding, respectively. Subareas 3.1 and 3.2 apply to several types of educational institutions at different levels, comprising primary, lower, and upper secondary education, as well as HE. For instance, such projects aim to reduce socio-economic inequalities of students belonging to lower socio-economic status at 400 selected primary or secondary schools in peripheral regions/locations. The aim is to reduce inequalities through multifaceted support in the form of tutoring individual students from poorer families, providing them direct or indirect pedagogical assistance, setting up information channels between schools and families, assuring study and psychological support, peer learning, and exchange of effective practices.

Subarea 5.1 is intended to support only public HEIs. The only competitive call in subarea 5.1 was for setting up research centers of excellence, made within the framework of the EXCELES[6] program in 2021 comprising five priorities. The priorities were infectious diseases and virology, oncology, neuroscience, metabolic and cardiovascular diseases, and socio-economic impacts of systemic risks. For each priority, one project was selected (MŠMT, 2021). Subarea 5.1 thus comprises targeted financial support within the EXCELES program for the following five projects:

- National Institute of Virology and Bacteriology (undertaken by the Institute of Organic Chemistry and Biochemistry of the Czech Academy of Sciences)
- National Institute for Cancer Research (undertaken by Charles University)
- National Institute for Neurological Research (St. Anne University Hospital, Brno, affiliated to Masaryk University)
- National Institute for Research of Metabolic and Cardiovascular Diseases (Institute for Clinical and Experimental Medicine) and
- National Institute for Research on Socio-economic Impacts of Diseases and Systemic Risks (abbreviated as SYRI). This is a virtual scientific hub that brings together about 150 researchers and experts from Masaryk University (coordinating institution), Charles University, and institutes of the Czech Academy of Sciences.

Each institute has set research priorities. For SYRI, the priorities include researching communication, behavior, and risks on social networks, social resilience, socio-economic inequalities, societal polarization and populism, law, and governance (SYRI, 2023). The connections and cooperation of top researchers with a background in different social science fields will proceed particularly through a virtual hub, with no need to set up additional infrastructure. SYRI should generate original knowledge of excellent scientific rigor and practical application, especially in public administration (including communication with the public), health, healthcare, and human lifestyle. SYRI also aspires to actively address government

[6] EXCELES is a program of support in Research & Development & Innovation, and the sole implementation tool for component No. 5.1 – Excellent Research and Development in the Health Sector of the National Recovery and Resilience Plan.

business strategies, the labor market situation, and civic poverty reduction. As a virtual scientific hub, SYRI funding has been set for 2022 to 2025 (SYRI, 2023). SYRI outputs are varied, ranging from research papers to news feeds and thematic workshops/roundtables.

Subarea 7.4 is also exclusive to public HEIs. The first call in this subarea, made in December 2023, is for the transformation of teaching and curricula in selected academic and vocation-oriented study programs. Such transformation should be toward developing green (energy-saving) skills, resilience, sustainability, and ecologic renovation, including the Green Deal agendas. The goals are to create new study programs (minimum of 20), update the existing study programs with new courses (minimum of 50), and create at least 20 new lifelong learning courses. To kick-start this and other priorities under REPowerEU, the EU has already allotted EU 0.14 billion in the form of advanced funding at the end of 2023 (MPO, 2023).

Recovery and Resilience Plan: Concrete Project Activities at Public HEIs

As suggested, public HEIs can participate in four subareas of the RRP. Subarea 5.1 resulted in the establishment of five centers of excellence, with one of those centered on social science issues (SYRI). The centers are governed by Charles University and Masaryk University which are the two most research-intensive Czech universities, along with selected institutes of the Czech Academy of Sciences. Subarea 7.4 develops activities that address goals of green transformation and sustainability, accentuating the creation of learning environments that deepen students' relevant skills and speed up workforce transformation (MŠMT, 2023). To this end, there are five key activities, namely, accreditation of new study programs, creation of new study courses, making of new Long Life Learning (LLL) courses, formulation of a national strategy on sustainability and green transformation of Czech HEIs (by the Ministry of Education) and forming strategic partnerships among HEIs. The total financial support distributed to public HEIs in the form of block grant funding is CZK 960 million (EUR 38.4 million). With the first call publicized in December 2023, at the time of writing (early 2024), it is too early to expect specific results on the green transformation of Czech public HEIs in this subarea.

However, more specific activities and concrete projects have been undertaken by Czech public HEI, particularly in subarea 3.2. As stated, the goals in this subarea are as follows:

- *Transformation of HE curricula*, and more specifically, digitalization of educational activities and study agendas, development of distance and online learning platforms and blended learning, creation of new, vocation-oriented study programs, and creation of new courses aimed at upskilling and reskilling.
- *Creation of new study programs in progressive fields*, more specifically, digital technologies, quantum technologies, micro/nanotechnologies, bio/eco-technologies, space technologies, autonomous mobility, alternative propulsion, circular economy, and green design.
- *Joint projects (involving all HEIs)*, and more specifically, improving recognition of study through micro-credentials, security of distance mode of study, and digitalization of study-related agendas and administration.

The financial support amounted to CZK 2.1 billion for curricular transformation, CZK 600 million for newly designed progressive study programs, and CZK 252 million for joint projects. The support was distributed as block grant funding to 26 public HEIs between 2022 and 2023, totaling CZK 3 billion (EUR 0.12 billion).

The transformation of HE curricula entailed 26 projects, with each public HEI undertaking one comprehensive project. To give a few examples, Masaryk University in Brno (MUNI) adopted the project "MUNI 3.2.1," the Prague University of Economics and Business (VŠE) initiated "Modernisation and Digitalisation of VŠE," and The Czech University of Life Sciences (CZU) started "Transformation of CZU towards New Forms of Study and Changing Labour Market Needs." Similarly, the University of South Bohemia in České Budějovice (JU) opted for "Adaption of JU to New Forms of Study with Regard to a Changing Labour Market Situation," and the University of Ostrava (OU) undertook the project "Flexible OU to Develop Professional Skills." Overall, with one project per HEI, the intention was to involve (most) institutional units top-down, starting from the rectorate and faculty leadership (dean's offices) down to institutes and departments.

Second, to boost *the creation of new, progressive study programs, several such programs were* funded at a particular public HEI. For instance, Brno University of Technology had four innovative programs supported, namely, "Environmental Engineering," "Chemical Technologies and Nanotechnologies," "Nuclear Energy," and "Advanced Automotive Engineering." Similarly, Czech Technical University in Prague (CTU) developed the programs "Quantum ICT" and "prg.AI Master." MUNI initiated the preparation of "Data Analytics programme," "Virology programme," and an interdisciplinary program "Applied Health Economics." In this respect, Charles University too had three new innovative programs in the pipeline, namely, "Bioinformatics and Computing Technologies," "Epidemiology and Public Health Management," and "Digital Social Sciences."

Thirdly, *joint projects* funded under subarea 3.2 entail the cooperation of representatives of all 26 public HEIs, forming three thematic networks. The network working toward "improving recognition of study through micro-credentials" is coordinated by Charles University. It aims to improve the accessibility and recognition of HE studies using micro-certificates and digitalizing educational activities and administrative tasks. More specifically, the activities undertaken by the network:

> ... focus on an analysis of the current situation concerning the micro-credentials use in Czech HE and on an analysis of relevant foreign experiences. The aim is to understand the national and foreign situation, including identifying the most suitable practices. Second, a new methodology for recognizing HE courses, including utilizing micro-certificates, will be developed. The methodology will also comprise the proposal of a unified, nationwide template of a micro-certificate and the unification of short-term HE study outputs using the ECTS credit system. Third, a joint online catalogue of courses at Czech public HEIs that fulfil micro-certificate requirements, including a unified verification of the results and a repository, will be made until 2024. (UHK, 2022)

Correspondingly, the other two thematic networks focus on secure distance forms of study and the digitalization of study-related agendas and administration, respectively. The former is coordinated by a team from Czech Technical University in Prague, whilst the latter network is led by experts from MUNI.

Finally, it is possible to track the 100 most funded institutions from RRP to the present based on RRP total funding expenditures per institution (by December 2023). The corresponding list shows Charles University in third place (total support of CZK 1.58 billion/EUR 64.6 million), MUNI in sixth (CZK 1.2 billion/EUR 49.3 million), Palacký University in Olomouc coming sixteenth (CZK 550 million/EUR 22.5 million). Czech Technical University in Prague narrowly missed the first quartile, coming 26th (CZK 338 million/EUR 13.8 million).

Other Notable Climate-Saving Activities

Besides funding from the RRP, some Czech public HEIs perform other activities geared toward climate change mitigation. Two cases are highlighted because of their significance and impact. First, the Czech University of Life Sciences in Prague has approached sustainable development systematically for several years and developed its "Sustainability Strategy 2030" in line with the United Nations (UN) Sustainable Development Goals.[7] In the strategy, CZU envisions itself as a leader (among Czech HEIs) in researching and promoting solutions for sustainable development by 2030, eliminating or mitigating adverse impacts on the environment connected with CZU's activities. CZU aims to reduce its carbon footprint considerably by 2030 and reach carbon neutrality in 2050 (to make CZU compliant with the Paris Agreement). Importantly, such goals are turned into concrete activities and achievements, also listed in CZU's annual Corporate Social Responsibility Report. For instance, CZU has

- created a *smart campus based on the digitalization* of CZU operations and management to enable monitoring and optimization of resources such as water, electricity, waste, and data and capture of security and health risks
- developed *waste prevention* also through water filtration plants, sorted waste containers (on every building floor) including e-waste (around the campus site), and compost bins

[7] Developed as part of the UN Agenda 2030. There are 17 sustainable development goals ranging from elimination of poverty by ensuring affordable and clean energy to making sustainable cities and communities.

- implemented an alternative mode of transport for business trips of CZU management in the form of the inter-university car-sharing service "Uniqway," which uses the latest Škoda Auto vehicle models that meet strict emission standards
- put in operation *solar power plants and heat pumps* in selected CZU buildings
- introduced *water-saving measures* by installing green roofs and walls, capturing and using rainwater, using gray water, and taking care of lawns. To save consumption, the default setting of a small amount of water for automatic urinal flushing or water-saving taps with installed faucet aerators is used (CZU, 2022).

These examples attest to CZU's proactivity in systematically developing and upgrading its infrastructure to be in line with climate change mitigation strategies.

The second case is the UNILEAD project, standing for "University Leaders in Sustainable Development Goals," undertaken by a consortium of 24 Czech public HEIs coordinated by MUNI. The project aims to strengthen the role of universities as "efficient, responsible and inclusive" public organizations, by ensuring more effective cooperation in the transfer of good practice in the implementation of sustainable development goals within the operational aspects of universities (the so-called "Resources for Change"). Initiated in 2022, the project has the following ten key crosscutting areas, each area representing one resource for change: strategic partnerships; waste prevention and recycling; energy management; green procurements; sustainable cooking; sustainable mobility of staff and students; water management; green construction; biodiversity and sustainable IT infrastructure and electrization.

Each area is coordinated by the representatives of one HEI. For example, the expertise of the Czech University of Life Sciences in Prague in sustainable development (see above) is applied in the area of water management. Next, Palacký University in Olomouc coordinates sustainable mobility. Working on this priority, the project team researched the most common means of transport in the form of a pilot survey among Palacký University staff and students. The results revealed an untapped potential for bicycles on condition of the availability of the infrastructure, including safe paths and crossings and lockable on-site parking slots. The overall UNILEAD coordination is vested in a MUNI team that also leads activities under green procurements. Notably, developing such activities is in line with a sustainable strategy of MUNI as a part of the MUNI Strategic Plan 2021–2028. To keep developing environmental sustainability institution-wide, MUNI also set up a Sustainability Board, composed of representatives of each faculty and the rectorate, and started issuing a Sustainability Report available online annually.

Discussing Green Transition of Czech Universities

The analytical review of policies, plans, and practices for climate change mitigation by Czech public HEIs as well as state administration has yielded several findings worth discussing. First, factually, climate change mitigation and green transition are new agendas to the Czech HE sector and, to an extent, to some public administration units too (like the Ministry of Education). Although climate change and its thematically relevant issues figure in the overarching "Strategic Framework: Czech Republic 2030," there is little to no mention of them in strategic policy documents specific to HE. In this respect, the EU-level Green Deal clearly turned out to be a driver signaling the significance of the climate change mitigation policies to the Czech HE sector too. This was done especially through the "last-minute" inclusion of the REPowerEU priority with HE-specific subarea 7.4 "Adapting HEIs—promoting green skills and sustainability in HE" in the national RRP. With the effects of the REPowerEU priority yet to be ascertained, for now, it is safe to say that climate change is not a priority for HE policymaking *system-wise*, with the initiative so far more with individual HEIs and their clusters.

Second, in conceptual terms, the pivotal instrument for conveying the Green Deal climate change agendas has been the national RRP. More specifically, with its total allocation exceeding EUR 7 billion, the RRP is clearly a treasure type of agenda-setting instrument. It has been applied by the Czech government to foster demand, produce desired outputs, and initiate commissioned research (specifically in subarea 5.1). These characteristics have made the RRP pivotal, generating reactive action by public HEIs in the form of drawing from available funding (mostly block grants). To function, the RRP has a dedicated website (https://www.planobnovycr.cz/), with the priorities and calls under the managerial and administrative responsibility of the Ministry of Education having a specially dedicated website (https://www.edu.cz/npo/). These websites serve as instrument "nodes" for providing, spreading, and updating relevant information on RRP implementation.

Currently, RRP funding is still ongoing; so, no definite conclusions can be reached as to the RRP's effectiveness. However, the fact that no co-funding is needed provides good grounds for it. However, the assumed RRP effectiveness requires qualification regarding the relevance of (sub) priorities set for climate change mitigation. In this respect, the bulk of the funding is for curricular transformation and updates, including the creation of new courses and programs or progressing issues like micro-certificates, digitalization of study agendas, study administration, and so on. Although HE curricular modernization is part of the green transition, there has been very little beyond it in individual calls made so far in subareas 3.1, 3.2, 5.1, and 7.1. No specific attention seems to have been paid, for example, to circular economy-based solutions, smart means of transport, or energy-saving solutions and technologies at HEIs, including large-scale infrastructural investments. Support specific to medical research in HE-focused subarea 5.1 is also a bit odd. It is probably a (post-) COVID-19 undertaking with big potential for the future, but surely, other disciplines would have also been eligible (beyond social science-based SYRI).

To the author's best knowledge, no nation-wide debates have been held on such issues, open to the HE community and the experts. This is limiting because there have been no overarching frameworks mapping standpoints, connections, overlaps, and, importantly, the order of priority among the variety of activities that lay claim to saving the climate (also comprising funding allocations and specification of implementation measures). Should HEIs concentrate on building modernization? Or on extending the life cycle of institutional devices through the circular economy? On smart digitalization of devices? Or rather on smart means of transport? Should there be no priority whatsoever? These questions remain unanswered for now. Correspondingly, *the RRP shows little support for HE-related climate-saving activities per se*, with the focus placed on ancillary activities represented by phasing in new study courses or programs, curricular transformation, ICT, or biomedical research. Useful as these activities are, they could be productively funded from other sources. Perhaps the fact that Czech HEIs "are not jubilant" over the green agenda has something to do with a specific organizational response that sees climate change simply as just another agenda to be coped with under a condition of HEIs (and academics) constantly facing a capacity and mission overload (Benneworth et al., 2016; Pinheiro & Stensaker, 2013).

With the RRP providing and fostering rather reactive actions, Czech public HEIs undertake other activities falling under the green agenda.

An example is the UNILEAD project developed on a network basis by a consortium of Czech public HEIs in areas such as green procurements, sustainable mobility of staff and students, energy management, and waste prevention that the RRP does not cover (yet). However, with few exceptions, the UNILEAD outputs are mostly "on paper only" reviews and recommendations. Still, there are also some examples of pro-active institutional-level action and measures in climate change mitigation adopted by individual HEIs, namely CZU as well as MUNI.

The foregoing debate "on Green-dealing" and other climate-saving activities in Czech HE generates four pointers of international relevance. Firstly, the reliance on one pivotal instrument (RRP) attests to a single-tool policy design. This can be seen as an untypical, if not obsolete, way of designing sectoral policies these days, as public policy designs have for some time shown a proclivity to the application of several tools in combination (cf. Howlett, 2024). Secondly, the more pro-active versus more reactive climate engagement of Czech HEIs resulting in the institutional divide attests to the existing evidence from abroad, where HEIs display similar divisive tendencies. Thus, for now, corporate social responsibility for climate change mitigation is not taken for granted in the modern university (cf. Jongbloed et al., 2008). Thirdly, the scope of climate-saving activities performed by Czech HEIs is broad, ranging from new thematic courses/programs through infrastructural modifications to specialized research, which again corresponds to the bewildering variety of HEIs' actions in other countries. In these respects, there is clearly a ground for convergence as a governance mechanism of harmonization of global HE green-policy agendas (see Chap. 1, this volume). However, it would take much more specific ethnographic research to draw more definite conclusions about their actual "on/the ground" effects and impacts on the green transition in Europe and beyond, which is what matters ultimately (Friberg, 2016). However, it may well be that, recent mass protests by farmers and other front-level stakeholders against the Green Deal in Europe will force a political policy response at the EU level, regardless of any HE involvement.

Conclusion

The green agenda, including the Green Deal goals, is still rather new to Czech HE, with limited evidence available. However, Czech public HEIs have some funding at their disposal, including but not limited to the

RRP. Future research should concentrate on the mid-term or ex-post analysis of the RRP's impact as a major systemic policy tool set for the green agenda. The same goes for other tools like the UNILEAD. Such analyses reiterate the need to pay specific attention to the logic (or the lack of it) of a combination of individual policy tools into policy mixes. The policy mix formation should, in turn, help one penetrate the dynamics and complexity of climate change policy implementation (as highlighted in Chap. 1). Correspondingly, assessments of relevant policy drivers should provide insight into the political and policy intricacies of green agenda-setting. The assessments can be done comparatively within the Visegrad countries (Poland, the CR, Slovakia, and Hungary) or Europe-wide. At the institutional level, the preference should be for inquiries into the approaches of leading public HEIs to gain examples of good practice, as well as into the situation of small private Czech HEIs getting effectively no external green-policy support.

Lastly, but importantly, there is a politico-societal note in the order. It is possible to fund climate-saving activities of public HEIs and other public sector organizations with billions of CZK, requiring no co-funding. However, it is a very short-sighted government policy to simultaneously impose climate-related burdens on the Czech populace. This is primarily done now by substantially rising household—and industry—electricity prices[8] or by requiring a compulsory replacement of solid-fuel home furnaces not adhering to emission standards I and II. With the replacement effective from September 2024, the estimates of such furnaces yet to be replaced[9] vary between 150,000 and 250,000 (Onergetice. cz, 2024). Such policies only "add fuel to the fire," running the danger of enhancing populist views of universities as privileged ivory towers and enticing "blame it on the evil Brussels" sentiments. Hence, the Green-dealing of Czech HE had better not proceed without broader societal considerations and remedial actions putting individual (vulnerable) citizens in the spotlight. After all, a socially just transition and an active social dialogue are the qualities the Green Deal does advocate for.

[8] It is an open secret that electricity generated in Czech power plants is made available long-term to the electricity grid in the neighbouring countries (especially Germany) at very competitive spot market prices.

[9] There is a subsidy programme available but administratively still rather complicated and tied up in bureaucratic rules. Moreover, the residents concerned are often the elderly, who may not be well versed in the bureaucratic jargon underpinning a subsidy application.

Annex

Table 7.2 RRP areas, components, and associated funding

Number	*Component name*	*Funding (EUR million)*
Digital transformation		
1.	1.1 Digital services for citizens and businesses	112
2.	1.2 Digital government systems	277
3.	1.3 Digital high-capacity networks	227
4.	1.4 Digital economy and society, innovative start-ups, and new technologies	224
5.	1.5 Digital transformation of enterprises	196
6.	1.6 Acceleration and digitalization of the building permit process	57
Physical infrastructure and green transition		
7.	2.1 Sustainable and safe transport	943
8.	2.2 Reduction of energy consumption in the public sector	325
9.	2.3 Transition to cleaner energy sources	262
10.	2.4 Developing clean mobility	192
11.	2.5 Building renovation and air protection	632
12.	2.6 Nature protection and climate change adaptation	542
13.	2.7 Circular economy and recycling and industrial water	141
14.	2.8 Brownfields	131
15.	2.9 Promotion of biodiversity and drought issues	117
Education and labor market		
16.	**3.1 Innovation in education in the context of digitalization**	**191**
17.	**3.2 Adaptation of the capacity and orientation of school programs**	**517**
18.	3.3 Modernization of employment services and labor marketdevelopment	886
Institutions, regulation, and business support		
19.	4.2 New quasi-equity instruments to support entrepreneurship	39
20.	4.3 Anti-corruption measures	
21.	4.4 Enhancing the efficiency of the public administration	1
22.	4.5 Developing the cultural and creative sectors	214
Research, development and innovation		
23.	**5.1 Excellent R&D in priority areas of public interest in healthcare**	**196**
24.	5.2 Support for R&D in companies and the introduction of innovations into business practice	126

(*continued*)

Table 7.2 (continued)

Number	*Component name*	*Funding (EUR million)*
Health and resilience of population		
25.	6.1 Increasing health system resilience	153
26.	6.2 National plan to strengthen oncological prevention and care	335
REPowerEU (additional priority)		
27.	7.1 Renewable energy infrastructure and electricity grid	371
28.	7.2 Promoting decentralization and digitalization of the energy sector	3.2
29.	7.3 Comprehensive reform of the renovation wave consultancy in the Czech Republic	160
30.	**7.4 Adapting HEIs—promoting green skills and sustainability in higher education**	**40**
31.	7.5 Decarbonization of road transport	30.4
32.	7.6 Electrification of rail transport	52.7
33.	7.7 Simplifying environmental permission procedures and identifying areas for renewable energy development	12.4
Total		7706

Note: Grayed areas under the management and administration of the Ministry of Education

Adopted from EC (2021), Analysis of the Recovery and Resilience Plan of Czechia

References

Alves, F., Filho, W. L., Casaleiro, P., Nagy, G. J., Diaz, H., Al-Amin, A. Q., et al. (2020). Climate change policies and agendas: Facing implementation challenges and guiding responses. *Environmental Science & Policy, 104*, 190–198.

Bali, A., & Halpin, D. (2021). Agenda-setting instruments: Means and strategies for the management of policy demands. *Policy and Society, 40*(3), 333–344.

Benneworth, P., Pinheiro, R., & Sánchez-Barrioluengo, M. (2016). One size does not fit all! New perspectives on the university in the social knowledge economy. *Science and Public Policy, 43*(6), 731–735.

Cairney, P. (2012). *Understanding public policy: Theories and issues* (1st ed.). Palgrave Macmillan.

Centrum pro stadium vysokého školství (CSVŠ). (2018). *Higher education in the Czech Republic 2018*. CSVŠ.

Český statistický úřad (ČSÚ). (2022). *Vysoké školy v Česku studuje cca 300 tisíc studentů, klesá zájem o techniku, v kurzu je zdravotnictví (About 300,000 students are studying at universities in the Czech Republic, interest in technology is declining, and the course is healthcare)*. ČSÚ. https://www.czso.cz/csu/czso/

vysoke-skoly-v-cesku-studuje-cca-300-tisic-studentu-klesa-zajem-o-techniku-v-kurzu-je-zdravotnictvi

Cobb, R. W., & Ross, M. H. (Eds.). (1997). *Cultural strategies of agenda Denial: Avoidance, attack and redefinition*. University Press of Kansas.

CZU. (2022). Corporate and social responsibility report. https://csr.czu.cz/en/r-15346-strategy-and-sustainability-report

Dlouhá, J., & Dlouhý, J. (2014). Higher education for sustainability – a change of education genre? *Envigogika, 9*(1), 1–31.

Dlouhá, J., & Pospíšilová, M. (2018). Education for Sustainable Development Goals in public debate: The importance of participatory research in reflecting and supporting the consultation process in developing a vision for Czech education. *Journal of Cleaner Production, 172*, 4314–4327.

EC. (2021). Czechia's recovery and resilience plan. European Commission. https://commission.europa.eu/business-economy-euro/economic-recovery/recovery-and-resilience-facility/country-pages/czechias-recovery-and-resilience-plan_en#whats-in-the-plan

EU. (2019). Communication from the Commission to the European Parliament, the European Council, the Council, the European Economic and Social Committee and the Committee of the Regions. The European Green Deal. Brussels: European Commission https://eur-lex.europa.eu/resource.html?uri=cellar:b828d165-1c22-11ea-8c1f-01aa75ed71a1.0002.02/DOC_1&format=PDF

European Parliamentary Research Service (EPRS). (2021). *Climate action in Czechia – Latest state of play*. European Union.

Filho, W. L. (Ed.). (2010). *Universities and climate change: Introducing climate change to university programmes*. Springer.

Filho, W. L., Eustachio, J. H. P. P., Caldana, A. C. F., Will, M., Salvia, A. L., Rampasso, I. S., Anholon, R., Platje, J., & Kovaleva, M. (2020). Sustainability leadership in higher education institutions: An overview of challenges. *Sustainability, 12*(9), 3761–3780.

Filho, W. L., Shiel, C., Paço, A., Mifsud, M., Ávila, L. V., Brandli, L. L., Molthan-Hill, P., Pace, P., Azeiteiro, U. M., Vargas, V. R., & Caeiro, S. (2019). Sustainable Development Goals and sustainability teaching at universities: Falling behind or getting ahead of the pack? *Journal of Cleaner Production, 232*, 285–294.

Friberg, T. (2016). Generating ethnographic research questions: An anthropological contribution to the study of higher education. *Ethnography and Education, 11*(3), 345–358.

Government of the Czech Republic (CR Government). (2012). *The National Priorities of Oriented Research, Experimental Development, and Innovations*. Úřad vlády ČR. http://www.czech-research.com/wp-content/uploads/2016/09/National-priorities-of-oriented-research-experimental-development-and-innovations.pdf

Green, A. H., & Cohen, D. K. (2021). Centering human subjects: The Ethics of "desk research" on political violence. *Journal of Global Security Studies, 6*(2). https://doi.org/10.1093/jogss/ogaa029

Hainsch, K., Löffler, K., Burandt, T., Auer, H., del Granado, P. C., Pisciella, P., & Zwikl-Berndhard. (2022). Energy transition scenarios: What policies, societal attitudes, and technology developments will realize the EU Green Deal? *Energy, 239*, 122067.

Hood, C. (1986). *The tools of government*. Chatham House Publishers.

Howlett, M. (1997). Issue-attention and punctuated equilibria models reconsidered. An empirical examination of the dynamics of agenda setting in Canada. *Canadian Journal of Political Science, 30*(1), 3–29.

Howlett, M. (2024). *Designing public policies: Principles and instruments* (3rd. ed.). Routledge.

Jongbloed, B., Enders, J., & Salerno, C. (2008). Higher education and its communities: Interconnections, Interdependencies and a research agenda. *Higher Education, 56*(3), 303–324.

Levin, K., Cashore, B., Bernstein, S., & Auld, G. (2012). Overcoming the tragedy of super wicked problems: Constraining our future selves to ameliorate global climate change. *Policy Sciences, 45*(2), 123–152.

Marek, M., & Krejza, Z. (2023). Circular economy development both in the Czech Republic and the world. *Procedia Computer Science, 219*, 1678–1686.

McCowan, T. (2019). *Higher education for and beyond the sustainable development goals*. Palgrave Macmillan.

McCowan, T. (2020). *The impact of universities on climate change: A theoretical framework*. Climate-U, Institute of Education, UCL.

McCowan, T., Filho, W. L., & Brandli, L. (Eds.). (2021). *Universities facing climate change and sustainability*. Körber-Stiftung.

Meyer-Sahling, J. H. (2011). The durability of EU civil service policy in Central and Eastern Europe after accession. *Governance, 24*(2), 231–260.

Ministerstvo průmyslu a obchodu (MPO). (2021) Národní plan obnovy – hlavní pilíře [Recovery and resilience plan – main pillars]. https://www.planobnovycr.cz/hlavni-pilire

Ministerstvo průmyslu a obchodu (MPO). (2023). Aktualizace Národního plánu obnovy [Update to recovery and resilience plan]. https://www.planobnovycr.cz/aktualizace-npo

Ministerstvo školství, mládeže a tělovýchovy (MŠMT). (2021). Program EXCELES – Implementační nástroj komponenty 5.1 Národního plánu obnovy. https://www.msmt.cz/vyzkum-a-vyvoj-2/program-exceles-komponenta-5-1-narodniho-planu-obnovy

Ministerstvo školství, mládeže a tělovýchovy (MŠMT). (2023). *Ministry of Education, Youth and Sports*. Online at: https://msmt.gov.cz/

Ministry of Education, Youth and Sports (MEYS). (2020). *The strategic plan of the Ministry for higher education for the period from 2021*. MEYS.

O Energetice.cz. (2024). Asociace: Letos bude nutné vyměnit stovky kotlů, kapacity topenářů nestačí (Association: Hundreds of boilers will have to be replaced this year, the capacity of heating operators is not enough) https://oenergetice.cz/teplarenstvi/asociace-letos-bude-nutne-vymenit-stovky-tisic-kotlu-kapacity-topenaru-nestaci

Páleníková, M., Špetík, O., Páleník, R., Ilík, J., & Adamík, M. (2022). Implementation of sustainable public procurement in the rail transport in the Czech Republic and a comparison with the other members of the Visegrad Group. *NISPAcee Journal of Public Administration and Policy, 15*(1), 167–199.

Pinheiro, R., & Stensaker, B. (2013). Strategic actor-hood and internal transfromation: The rise of the 'Quadruple-Helix University'? In J. Branković, M. Klemenčič, P. Lazetić, & P. Zgaga (Eds.), *Global challenges, local responses in higher education* (pp. 171–189). Sense Publishers.

Plaček, M., Valentinov, V., del Campo, C., Vaceková, G., Ochrana, F., & Šumpíková, M. (2021). Stewardship and administrative capacity in green public procurement in the Czech Republic: Evidence from a Large-N survey. *Environmental Sciences Europe, 33*(94). https://doi.org/10.1186/s12302-021-00534-7

Rittel, H. W., & Webber, M. M. (1973). Dilemmas in a general theory of planning. *Policy Sciences, 4*(2), 155–169.

Shivakoti, R., & Howlett, M. (2022). Agenda-setting tools in theory and practice. In M. Howlett (Ed.), *The Routledge handbook of policy tools* (pp. 60–72). Routledge.

Sila, U., & Frohm, E. (2023). *Towards net zero in the Czech Republic* (p. 1754). OECD Economics Department Working Papers no.

Siuda, V. (2023). Green transition: Identifying vulnerable industries and bank loans in the Czech Republic. *Eastern European Economics, 62*(6), 679–693. https://doi.org/10.1080/00128775.2023.2204841

Slavík, J., Remr, J., & Vejchodská, E. (2018). Relevance of selected measures in transition to a circular economy: The case of the Czech Republic. *Detritus, 1*(1), 144–154.

SYRI (2023). Research. https://www.syri.institute/research

UHK. (2022). Specifický cíl C1: Zlepšení prostupnosti vzdělání na úrovni vysokých škol pomocí mikrocertifikátů (Specific objective C1: Improving the accessibility of education at the university level using micro-certificates). https://www.uhk.cz/cs/univerzita-hradec-kralove/veda-a-vyzkum/programy-projekty-a-souteze/narodni-plan-obnovy-2022-2024/h

Weiss, J. A. (1989). The powers of problem definition: The case of government paperwork. *Policy Sciences, 22*(2), 97–121.

CHAPTER 8

Responses to Climate Change: Context, Action, and Future Directions of China's Universities

Huang Yu, Tao Linlin, and He Botao

Introduction

Climate change is a significant global issue that affects all countries universally. China, the largest developing nation globally, has implemented various policies, measures, and activities to combat climate change. China has actively engaged in global climate governance and has achieved favourable outcomes in addressing this issue. Chinese government figures reveal that in 2020, China successfully reduced its carbon emission intensity by 18.8% compared to 2015, exceeding the target set in the 13th Five-Year Plan 2016–2020. Additionally, the figure was 48.4% lower than the value in 2005, indicating that China has successfully met its obligation to the global community by achieving a decrease in carbon intensity of 40–45% from the 2005 level by 2020. China's efforts substantially reduced

H. Yu (✉) • T. Linlin • H. Botao
Institute of International and Comparative Education, Beijing Normal University, Beijing, China
e-mail: huangyu@bnu.edu.cn; taolinlin@mail.bnu.edu.cn; hebotaoi@163.com

J. Kekäle, R. Pinheiro (eds.), *Higher Education Policy for Tackling Climate Change*, https://doi.org/10.1007/978-3-031-78479-8_8

approximately 5.8 billion tonnes of carbon dioxide emissions between 2005 and 2020, showcasing a significant decrease in carbon emission intensity (Chinese Government, 2021). China experienced an average annual increase in energy consumption of 3% from 2012 to 2021, which supported the country's 6.6% average economic growth rate (Chinese State Council, 2022). Over this time frame, there was a 34.4% decrease in carbon dioxide emissions for each unit of gross domestic product (GDP), and a 26.4% decrease in energy consumption for each unit. This reduction is equivalent to a decrease of 1.4 billion tonnes of standard coal (IREA, 2021).

Universities play a crucial role in both adapting to and mitigating climate change. They actively fulfil their responsibilities and provide invaluable contributions to tackling climate change. The higher education sector in China has implemented several initiatives to support national and international plans aimed to address climate change. These measures primarily focus on scientific research, talent development, and international collaborations. Several colleges have included elements of the Sustainable Development Goals (SDGs) into their curriculum design to equip students with the essential information, skills, and values. Several institutions have established graduate programmes, including master's and doctoral degrees, focusing on sustainable development to educate and prepare professionals in this field. In addition, certain universities have developed research organizations focused on sustainable development and have collaboratively established worldwide collaboration platforms to spearhead research on climate change.

Academics, both domestically and internationally, have examined it from various viewpoints. The analysis examined campus operations, talent development, scientific research, and social service. Feng Lei (2022) emphasizes the significance of these areas in addressing climate change. Fankhauser et al. (2023) discussed the actions universities can implement to tackle climate change across five key areas: education, scientific investigation, community involvement, reduction of campus emissions, and financial investments. Tang and Wang (2024) conducted a case study on the School of Life Sciences at Capital Normal University to investigate the social practices and models of climate change education. They examined four aspects: background, measurements, efficacy, and experience. Previous studies contribute a macro-analytic framework to showcase the endeavours of Chinese universities in tackling climate change. Drawing upon the research of these scholars, it examines three key aspects: China's national policies, the primary actions taken by Chinese universities to combat climate change, and the models employed by Chinese universities in addressing this issue.

Context: Active National Strategies to Address Climate Change

While colleges have some autonomy in dealing with climate change, their actions are guided mainly by national policy. Hence, understanding China's overarching strategy for climate change legislation is essential for understanding Chinese universities' unique activities and behavioural patterns in response to climate change.

In response to the Bali Action Plan, the Chinese government established the National Leading Group for Addressing Climate Change, Energy Conservation, and Emission Reduction in 2007 (China State Council, 2007). The National Climate Change Programme was launched in the same year, outlining the specific steps necessary to tackle the problem of climate change, which marked the first step in implementing policies and practices to adapt to climate change (Wang H. C., & Wang Y. J., 2022). During the period of the "12th Five-Year Plan," the Chinese government prioritized the issue of climate change. China developed its first National Climate Change Adaptation Strategy in 2013. This strategy presented a thorough summary of the essential prerequisites, critical steps, regional discrepancies, and preventive strategies for adapting to climate change from 2014 to 2020. In addition, it offered direction and a foundation for adjusting to climate change (MOEE et al., 2022). The China–U.S. Joint Announcement on Climate Change was released during the APEC Beijing Summit 2014. The proclamation outlines the specific goals for further action beyond 2020 established by the United States and China. China has formally declared its carbon emissions will reach their highest point by 2030. China claims that the percentage of non-fossil energy in primary energy will rise to 20% by 2030 (Information Office of State Council, 2014). The Chinese government formulated the National Plan on Climate Change (2014–2020) to delineate the necessary actions to be taken by 2020 to advance climate change adaptation. This plan has played a crucial role in promoting efforts to address climate change (Information Office of State Council, 2014).

During the UN Climate Change Conference in Paris in December 2015, Chinese President Xi emphasized the importance of creating a just and efficient global framework to tackle climate change. Furthermore, he declared China's objective of decreasing its carbon emissions per unit of GDP by 40%–45% by 2020, compared to the levels recorded in 2005. China officially joined the Paris Agreement on September 3, 2016, after receiving approval from the Standing Committee of the National People's Congress (Zhou, D. C., & Zhang, L. X., 2016). Consequently, China joined the 23 parties that had formally endorsed the agreement. In September 2020, Chinese President Xi Jinping announced at the General Debate of the 75th Session of the United Nations General Assembly that China aims to peak its carbon dioxide (CO_2) emissions before 2030 and achieve carbon neutrality by 2060 by implementing robust policies and measures (Chinese Government, 2020). During the Climate Ambition Summit on December 12, 2020, President Xi Jinping announced that China aims to decrease its carbon dioxide emissions per unit of GDP by almost 65% by 2030, compared to the levels recorded in 2005. Furthermore, China aims to guarantee that non-fossil energy sources account for 25% of its primary energy consumption (Xi, J. P., 2020).

The Ninth Meeting of the Central Committee for Financial and Economic Affairs in March 2021 placed significant importance on integrating the goals of achieving the highest level of carbon emissions reduction and attaining carbon neutrality into the broader strategy for constructing an ecological society. Furthermore, diverse ministries, central enterprises, provinces, and municipalities implemented numerous pertinent policies and regulations. The Ministry of Ecology and Environment has created the Guiding Opinions on Coordinating and Strengthening the Efforts in Climate Change and Ecological Environmental Protection, together with the Administrative Measures for Carbon Emissions Trading (for Trial Implementation), as illustrations. Furthermore, the State-owned Assets Supervision and Administration Commission (SASAC) has revised the Measures for the Supervision and Administration of Energy Conservation and Ecological Environment of Central Enterprises (Draft for Comments) (Chu, Z. W., 2021). The Ministry of Science and Technology has spearheaded the development of the Technology Lists for Addressing Climate Change (Wang et al., 2021a). In July 2021, the Ministry of Education released the Action Plan for Carbon Neutralisation and Scientific and Technological Innovation in Colleges and Universities. This approach clearly defines the primary objective of colleges and universities, which is to offer technological assistance and develop skilled individuals to achieve carbon peaking and neutrality (Wang et al., 2021b). This undertaking is a proactive measure that aligns with the national strategy.

Furthermore, some non-governmental organizations are actively implementing measures. The China Youth Climate Action Network (CYCAN), established in August 2007, seeks to encourage young individuals to gain information, comprehend, and actively engage in activities to tackle climate change, typically through advocating for specific actions, implementing practical measures, fostering international partnerships, and promoting environmentally friendly job opportunities (Zhang, Z. G., 2017). Around 500 Chinese institutions are participating in the project established by CYCAN (Jiang, L., 2019). The "Low Carbon Campus" Project, initiated by CYCAN in 2007, has garnered participation from students from more than 50 Chinese colleges, with an approximate annual attendance of 1000 students (Li, P. H., 2016).

China has made significant climate change management progress since implementing the 12th Five-Year Plan. The nation has accomplished its goal of mitigating climate change by 2020 ahead of schedule and has observed positive outcomes in reducing pollution and advancing the economy. During the latter part of the 13th Five-Year Plan Period, it substantially restructured its climate change organizations and management institutions, introducing essential modifications to enhance effectiveness. Both national and regional climate policies and activities have been revised to prioritize the cooperation of climate policies and environmental policies (Zhu & Zhu, 2020).

Although the Chinese government has achieved positive results, China remains the world's largest carbon emitter. Therefore, China consistently reviews and refines its policies to establish a robust foundation for future initiatives and advancements. Information Office of China State Council published a White Paper titled "Responding to Climate Change: China's Policies and Actions" in October 2021 (Chinese Government, 2021). The document offers a concise overview of China's endeavours and lessons learned in adjusting to climate change, with the additional objective of fostering global comprehension. The National Strategy for Climate Change Adaptation 2035 was jointly issued in June 2022 by the Ministry of Ecology and Environment, and 17 other Chinese agencies (MOEE et al., 2022). This strategy establishes a target for 2035, which was defined through a comprehensive assessment of the hazards presented by climate change and the efficacy of climate change adaptation measures. The strategy, moreover, delineates the pivotal domains, geographical trends, and precautionary actions imperative for China's adjustment to climate change.

In December 2023, the Ministry of Ecology and Environment published the 2023 Annual Report on Climate Change Response (MOEE, 2023). This Annual Report focuses on China's latest policy efforts to address climate change. The yearly report highlights China's active efforts in minimizing climate change and its proactive adaptation approach. In the next decade, China will prioritize addressing climate change in its governance, enhance efforts to respond to climate change, promote a comprehensive shift towards environmentally friendly economic and social development, and strive to achieve a harmonious coexistence between humans and nature through modernization. In February 2024, The State Council issued the Provisional Regulations for the Management of Trade in Carbon Emission. The rule is the first specialized regulation in China's climate change domain, and it is the first instance where the carbon emission trading system has been formally defined by administrative regulations, symbolizing a significant milestone.

Actions: Closely Intertwined Implementation of the Fundamental Activities of the University

In contemporary society, higher education institutions assume a pivotal role and hold a distinct position. Universities have the crucial responsibility of nurturing new minds and offering intellectual assistance to society. Universities are the optimal setting for constructing a sustainable development framework and executing strategies to tackle climate change (Yang, K., 2009). Chinese universities have made meaningful contributions to China's combating climate change and have effectively supported the implementation of key government policies and programs. Particularly with the introduction of the "carbon peaking and neutrality" objective, several schools and institutions have implemented measures to address this issue. The Yangtze River Delta Sustainable Development University Alliance was formed in April 2021 at the suggestion of Tongji University (Chu, Z. W., 2021). The Alliance is a collaborative effort established by eight prestigious universities: Southeast University, Fudan University, East China Normal University, Nanjing University, Shanghai Jiaotong University, Tongji University, Zhejiang University, and the University of

Science and Technology of China. The Alliance, the pioneering university alliance in China dedicated to achieving carbon neutrality, has released the Action Initiative for Advancing Carbon Peaking and Carbon Neutrality in Universities. Additionally, it has endorsed the Charter of the Yangtze River Delta Sustainable Development University Alliance. Several other colleges and universities have implemented campus development measures to comply with the Ministry of Education's mandate to create environmentally sustainable campuses. China University of Geosciences (Wuhan) has adopted the theme of "Building a Beautiful China and A Habitable Earth" for its strategic planning to become a globally recognized research university in the field of geosciences by 2030 (Wang et al., 2021b). As part of its commitment, the university aims to achieve carbon neutrality on its campus by 2052, coinciding with its centenary, and demonstrating its long-term dedication to sustainable development. Chinese universities' response to climate change is mostly focused on three areas: people training, scientific research, and international exchanges. These activities are closely aligned with the institutions' roles.

Course Teaching and Talent Development

The primary measures for fostering talent encompass two key components: establishing disciplinary frameworks and enhancing curriculum design. Disciplines on climate change in China have been incorporated into the Specialty Catalog of Undergraduate Courses for mainstream colleges and universities, focusing on discipline building. In 2019, a total of 87 climate change–related majors from 81 universities, including Peking University, Tsinghua University, Nanjing University, and Tianjin University, were designated as national first-class undergraduate majors under the "Double Ten Thousand Course Plan." This plan, aiming to establish 10,000 national first-class undergraduate majors and 10,000 provincial first-class undergraduate majors, is a significant initiative in talent development. Tongji University and Renmin University of China have established master's and doctorate programs focused on sustainable development (Guangming Daily, 2023). These programs aim to teach individuals who are urgently required to address pressing global issues such as the "energy crisis," "environmental pollution," and "climate change."

Chinese universities offer various courses related to curriculum development, but the emphasis of these courses differs. For example, courses like "An Introduction to Climate," "An Introduction to Meteorology," and "Modern Climatology" prioritize theoretical knowledge of climate meteorology. On the other hand, courses such as "Marine Environment and Global Climate Change" and "Climate Change and Human Society" focus on the impact of climate change on specific ecosystems. Additionally, some universities organize special lectures where experts from different fields give thematic reports. Examples include the doctoral course on "Global Climate Change" offered by the Institute of Atmospheric Physics of the Chinese Academy of Sciences and the Climate Change Global Lectures at Tsinghua University (Liao et al., 2022). In addition, Chinese colleges are aggressively launching online courses on climate change. The Ministry of Education has designated specific courses as the initial selection of national first-class undergraduate courses in 2020. These include "China's Perspective on Climate Change" at Tsinghua University, "Air Pollution Control" at Tianjin University, "Air Pollution Control Engineering" at Huazhong University of Science and Technology, and "Climate Change and Human Society" at Nanjing University of Information Science and Technology (Guo, R., & Ren, Z. R., 2022).

The most renowned and practical course in climate change is the "Introduction to Climate Change Science." This course is primarily taught by leading scientists from the Chinese Academy of Sciences, and the Chinese team of the Intergovernmental Panel on Climate Change (IPCC), along with other academicians and scholars actively involved in climate change research, provides a comprehensive understanding of the most recent advancements and key areas of concentration in climate change research from a global perspective. It covers the facts, causes, and forecasting of climate change, the consequences and adjustment to climate change, efforts to mitigate climate change and promote sustainable development, as well as the most recent advancements and developments in global climate change negotiations and China's actions to reduce emissions, along with future directions for progress. This course, with its global perspective, aims to equip students with a comprehensive understanding of the most recent advancements and critical areas of concentration in climate change research. It also aims to help students develop a firm grasp of climate change science, providing solid groundwork for future research. Since 2018, the course has primarily been taught as a general course at prestigious universities in China, including the University of Chinese Academy of Sciences, Peking University, Tsinghua University, Sun Yat-sen University, Nanjing University, Beijing Normal University, Lanzhou

University, Yunnan University, and Nanjing University of Information Science and Technology (Liao et al., 2022).

Nanjing University of Information Science and Technology (NUIST) is a Chinese university dedicated to developing its talent training system to develop skilled individuals who can effectively handle climate change. NUIST, which traces its roots back to the establishment of Nanjing Institute of Meteorology in 1960, is a specialized institution with a strong emphasis on meteorological sciences and an industrial orientation (NUIST, 2023). Throughout its history, NUIST has developed a distinctive method for nurturing talent, which will be explained in the following five characteristics. Initially, the objective is to implement a multiple selection mechanism to transition from a selective talent selection approach to an adaptive one (Jiang et al., 2020). This system entails recruiting exceptional students based on their personality traits that align well with the professional attributes of a university (Zhang, N. X., & Sun, J. H., 2021). Furthermore, the curriculum structure should be developed by considering the distinctive disciplines. NUIST, driven by the prestigious national first-class discipline of atmospheric sciences, fosters interdisciplinary collaboration and develops a curriculum structure to establish a distinguished discipline brand that combines meteorology with other fields (Li, B. Q., 2022). NUIST also established a cooperative training structure that included the government, industry, universities, and research institutions. NUIST strengthens the "trilateral co-construction" system with the Ministry of Education, the China Meteorological Administration, and the People's Government of Jiangsu Province. It expands the main co-construction body and establishes a diverse co-construction system involving "industry, school enterprises, school institutes, school sites, and international partners" (Li, B. Q., 2022). This system aims to gather high-quality educational resources from various sources, providing students with practical skills and competence in their field, making them feel prepared and capable (Wu, L. B., & Tian, Y. T., 2020).

Scientific Research and Knowledge Innovation

Chinese universities are assuming a progressively significant role in addressing climate change and are adopting an escalating number of responsibilities. In 2015, the Ministry of Education and the China Meteorological Administration collaborated to release the "Guiding Opinions on Strengthening the Training of Meteorological Talents." This document suggested improving the composition of meteorology-related academic programmes and increasing the number of undergraduate and graduate students specializing in atmospheric science, but in a controlled manner. Simultaneously, there was a proposal to establish a meteorological talent training alliance between the Ministry of Education and the China Meteorological Administration (Shao, L. L., 2020). This alliance aims to fully utilize the capabilities of the Steering Committee for the Teaching of Atmospheric Science and the National Meteorological Vocational Education Teaching Steering Committee in conducting research, providing consultation, offering guidance, evaluating, and serving undergraduate and vocational education in colleges and universities. Additionally, the alliance seeks to establish and enhance a collaborative innovation mechanism for meteorological science and technology. This mechanism will involve the participation of meteorological industry enterprises, universities, scientific research institutes, and local governments.

Chinese universities strategically utilize their disciplinary and geographical strengths to build research centres, think tank research institutions, and research teams dedicated to climate change research in many areas. Yunnan University is the only university in Southwest China authorized to offer a doctoral degree programme in "Atmospheric Sciences" and is recognized as the provincial leader in meteorology. Additionally, Yunnan University has formed the Yunnan University Provincial Innovation Team for Climate Change Research on the Greater Mekong Subregion, the sole research team in China dedicated to studying climate change in this region (Yang et al., 2022). The China Center for Climate Communication was established in April 2010 by the Research Center for Journalism and Social Development of Renmin University of China and Oxfam Hong Kong. It is the first institute in China dedicated to researching the theories and practices of climate change and climate communication (RMU, 2012). Shandong University Climate Change and Health Center was established in September 2010. In May 2011 (SDU, 2018), Fudan University established the Research Institute for the Changing Global Environment, leveraging its interdisciplinary strengths, to address significant scientific challenges in climate change science specifically (FDU, 2011).

Similarly, Xiamen University founded the Carbon Neutrality Innovation Research Center in December 2020, focusing on researching on marine carbon sequestration technology (Chu, Z. W., 2021). In May 2021, Northwest University collaborated with the local government to establish the Yulin Carbon Neutralization Science and Technology Innovation

Center. The primary objective of this initiative is to facilitate the transition of the local energy mix and industrial structure (Xue, B., 2014). In October 2022, China Cultural University and China Meteorological Administration collaborated to build a laboratory focused on studying the adaptability of agriculture to climate change (CAU, 2022). This partnership aims to leverage the unique strengths of both institutions to conduct cooperative research. Several universities collaborate with international institutions to develop research institutes crucial for advancing their fields of study. For example, Nanjing University of Information Science and Technology, leveraging its expertise in meteorological research, entered a contract with the Rock Ethics Institute at Pennsylvania State University (Xue, B., 2014). In October 2011, they established a think tank dedicated to climate change and public policy research (Jia et al., 2011).

Similarly, the UTSC-NWU Center for Climate Change Research was collaboratively established by Northwestern University and the University of Toronto Scarborough (UTSC) in April 2014 (Xue, B., 2014). Furthermore, in May 2021, Shanghai Jiaotong University established the Research Institute of Carbon Neutrality and Development and the China-UK Low Carbon College (Chu, Z. W., 2021). These institutions aim to develop a high-level think tank and conduct interdisciplinary research on low carbon and sustainable development.

Climate change science is continuously influenced by societal demands, technological advancements, and various other factors, shaping its development and direction. Climate science has evolved into a multidisciplinary field encompassing natural sciences, social sciences, and humanities, rather than solely focusing on pure natural science (Chen, D. L., 2019). As a result, domestic universities have implemented a multidisciplinary research approach focused on climate research. For instance, Peking University has explicitly stated its objective of nurturing climate talents, which involves developing individuals with proficient computer and foreign language abilities, a solid mathematical background, and a well-rounded cultural literacy (Li, B. Q., 2022).

Lanzhou University exemplifies multidisciplinary integrated research in the field of climate change. Lanzhou University is situated in a semi-arid environment, which gives it distinct regional features and a wide range of academic disciplines. In the year 2000, Lanzhou University was designated as the research platform for the "Key Laboratory of Western China's Environmental Systems, the Ministry of Education." This recognition was granted due to the university's expertise in physical geography, quaternary research, and atmospheric sciences. The platform studies climate change and environmental change specifically in Western China. From 2003 to 2007, Lanzhou University held the top position among all universities in China in terms of the overall quantity of international climate change research papers authored by Lanzhou University researchers (Li, W. L., & Dong, G. H., 2009). This accomplishment solidified Lanzhou University's significant contribution to climate change research in China. In 2021, two professors from Lanzhou University were recognized as among the top 1000 most prominent experts in climate change research (LZU, 2021). The Key Laboratory of Semi-Arid Climate Change, which operates under the Ministry of Education, has emerged as a prominent institution in climate change research. This system aims to predict the worldwide epidemic's development pattern accurately. The Key Laboratory of Semi-Arid Climate Change has garnered recognition for its outstanding quality, substantial growth potential, and considerable international impact, making significant contributions to climate change research. The laboratory aims to drive forward research on semi-arid climate change, cultivate a new generation of highly skilled professionals, and position China's semi-arid climate research at the forefront of global scientific progress (LZU, 2023).

International Exchanges and Cooperation

Chinese institutions frequently organize international exchange conferences to address climate change, leveraging their unique strengths. The focal points of these international contacts are crucial areas and cutting-edge subjects about climate change. In 2012, the Academy of Disaster Reduction and Emergency Management of Beijing Normal University collaborated with the National Drought Mitigation Center of the United States and other institutions to conduct research (Xiao, X., 2012). They also jointly organized an international conference on drought disasters called the International Conference on Drought Disaster Monitoring, Assessment, and Management in the Context of Global Climate Change. The School of Design at Hunan University, in collaboration with the Oslo School of Architecture and Design in Norway and the Cape Peninsula University of Technology in South Africa, organized the inaugural International Symposium on "Design Strategies for Climate Change" in May 2017 (Zhang, J., & Fan, K. X., 2017). The symposium aimed to foster discussions and knowledge sharing on the cutting-edge subject of sustainable design and manufacturing in the context of climate change.

The event brought together Chinese and foreign researchers who have achieved several consensuses on sustainable design and development, thereby fostering communication and cooperation between industries and institutions.

Renmin University of China is one of the universities actively participating in international exchanges (Zheng, B. W., & Zheng, Q., 2023). It is dedicated to enhancing China's international communication capacity and effectively addressing climate change issues. It began researching on climate communication after analysing the challenges faced and lessons the Chinese government, media, and NGOs learned during the Copenhagen Summit. China 4C assumes responsibility for "improving the narrative of China's response to climate change," "amplifying the Chinese voice," "effectively communicating China's response to climate change internationally," and "expediting the development of a climate communication framework with distinct Chinese characteristics." These responsibilities are based on the four pillars of "innovative ideas," "compelling storytelling," "enhanced capabilities," and "boosted self-confidence." China 4C has consistently prioritized amplifying the voice of adapting to global climate change over the past decade, while actively fostering climate communication in China and the international community. In addition, it has made significant contributions to theoretical study and the advancement of climate communication, particularly in international communication, resulting in sound good domestic and international effects. In the last decade, China's climate change and communication research has gained international recognition, as it has transitioned from being a participant and contributor to a leading role.

Recently, Tsinghua University, a prestigious Chinese university, has prioritized matters concerning climate change and has actively worked towards fostering international contacts and collaboration. In 2019, Tsinghua University spearheaded the formation of the Global Alliance of Universities on Climate (GAUC). GAUC serves as an international platform for collaboration, leveraging the strengths of universities in knowledge, ideas, culture, science, and technology to support grassroots efforts and take a leading role in addressing climate change on a global scale. Tsinghua University has established a prominent program focused on fostering debate and interaction in the field of climate that has a significant impact on a worldwide scale (THU, 2006). The program invites prominent global political leaders, heads of international organizations, renowned academics, and influential business leaders to deliver speeches and engage in dialogues focused on China's role and contribution in addressing climate change, promoting environmental sustainability, formulating policies, implementing global and European Union's climate actions, managing global climate governance, and strengthening the Sino–US climate relationship. The program also explores the involvement of youth in global climate change efforts and aims to foster discussions on national and global responses to climate change. Additionally, participants will exchange ideas on the necessary actions and initiatives to enhance global climate change governance. Tsinghua University collaborates with international authorities and organizations to conduct regional climate change initiatives. For example, the Asian Development Bank (ADB) and Tsinghua University collaborated to establish a regional knowledge hub focused on climate change to conduct pertinent research.

Future Direction: Enhance the Integration of University Initiatives with National Climate Change Strategies

Climate change is a globally significant topic that receives widespread attention. Climate change induces a series of interconnected negative consequences with direct and indirect impacts, presenting intricate difficulties that every nation must address. As the world's second-largest economy and a developing nation, China has proactively implemented measures to tackle climate change and engage in international climate governance. China has issued a range of policies, strategies, and action plans on climate change since the implementation of the "National Climate Change Program of China" in 2007. The Chinese government has integrated climate change initiatives into its national and departmental strategic planning processes. China consistently modifies its response plans to adapt to and mitigate climate change.

Universities have a vital role in the worldwide issue of climate change, and Chinese universities have made a substantial contribution to the progress of implementing climate change legislation in China. The conventional tripartite roles of universities, namely, fostering talent, conducting scientific research, and providing social services, are closely interconnected with efforts to address climate change. Several universities proactively grow climate-related fields of study, offering both online and offline courses on climate change, as well as master's and doctoral programs, to foster innovative and exceptional individuals in this area. Several universities have established climate change–focused research institutions, collaborating to create global cooperation platforms and spearhead cutting-edge climate change research. The actions of universities in response to climate change are closely linked to government policymaking, showcasing the characteristics as of Chinese universities aligning with national strategies and highlighting the harmonious collaboration between the Chinese government, departments, and universities.

Chinese universities have developed many approaches to address climate change, primarily through teaching, research, and international collaborations. Certain universities leverage their disciplinary strengths to develop innovative mechanisms for nurturing talent, devise distinctive approaches for cultivating climate change expertise, and actively engage in climate change governance. Several universities engage in climate change research, consistently enhancing their reputation in climate science research, and bolstering China's influence in the global climate academic community. Several institutions actively advocate for international exchanges and collaboration, establish sophisticated international cooperation platforms, and exemplify the notion of mutually beneficial cooperation in global climate initiatives. These examples illustrate the proactive measures Chinese universities took to tackle climate change. Chinese universities have implemented climate activities based on their assets, including disciplinary expertise, regional advantages, and worldwide influence. Their long-term nature and strategic planning characterize these initiatives. The aim is to position themselves at the forefront of climate leadership, actively engage in climate governance, and enhance their own and China's influence in climate matters. Chinese higher education institutions will maintain a crucial role in China's future efforts to address climate change. Universities will offer recommendations and make valuable contributions to the development, execution, and modification of national climate change policies through the fostering of talent, scientific research, and international collaborations. Additionally, they will contribute to international politics, economics, and scientific disciplines on climate change.

References

Chen, D. L. (2019, December 10). The interdisciplinary integration of meteorological science cares about the warmth and cold of human heart. http://jres2023.xhby.net/js/kj/201912/t20191210_6437946.shtml//陈德亮. 气象科学跨学科融合关心人类内心冷暖.

China State Council. (2007, June 3). Notice on the issuance of China's national program on climate change. Retrieved May 20, 2023, from https://www.gov.cn/govweb/gongbao/content/2007/content_678918.htm//国务院.关于印发中国应对气候变化国家方案的通知

Chinese Agriculture University (CAU). (2022, October 30). News and information. http://news.cau.edu.cn/art/2022/10/30/art_8769_887365.html//中国农业大学. 新闻资讯

Chinese Government. (2020, September 30). China sets hard targets for carbon reduction. Retrieved May 5, 2021, from https://www.gov.cn/xinwen/2020-09/30/content_5548478.htm //中国政府.减碳，中国设定硬指标

Chinese Government. (2021). White paper on responding to climate change in 2021: China's policies and action. Retrieved September 12, 2024, from https://www.gov.cn/zhengce/2021-10/27/content_5646697.htm //中国应对气候变化的政策与行动_白皮书_中国政府网

Chinese State Council. (2022). Foreign Ministry spokesperson: China has always been a doer and action player in tackling climate change. Retrieved September 12, 2024, from http://www.gov.cn/xinwen/2022-11/09/content_5725696.htm //国务院. 外交 部发言人：中国始终是应对气候变化的实干家和行动派

Chu, Z. W. (2021). The significance, strategy, and evaluation of universities' participation in carbon neutrality: Based on the concept of sustainable development. *阅江学刊, 13*(05): 61–71+121–122. //储昭卫.高校参与碳中和行动的意义、策略与评价——基于可持续发展理念.阅江学刊, 2021, 13(05): 61–71+121–122. https://doi.org/10.13878/j.cnki.yjxk.20210805.001

Fankhauser, S., Sugandha, S., Sundvor, I., Stephanie, H., & Gireesh, S. (2023). Net zero portfolio targets for development finance institutions: Challenges and solutions. *Abstract Global Policy, 14*(5), 716–729. https://doi.org/10.1111/gpol.v14.5; https://doi.org/10.1111/1758-5899.13286

Feng, L. (2022). Responsibility and cooperation: An analysis of British universities' comprehensive action to deal with climate crisis. 比较教育研究, (7): 49–57. //冯磊.责任与合作:英国大学全面应对气候危机行动探析.比较教育研究. 2022, (07): 49–57.

Fudan University (FDU). (2011). The Global Environmental Change Institute was officially established to connect with the forefront of global climate change research. 复旦学报(自然科学版), *50*(3): 281. //复旦大学. 全球环境变化研究所正式成立接轨全球气候变化研究前沿.复旦学报(自然科学版), 2011, 50(03): 281.

Guangming Daily. (2023, February 7). Universities should better contribute to the Sustainable Development Goals. Retrieved May 30, 2023, from https://m.gmw.cn/baijia/2023-02/07/36349774.html//光明日报. 大学应更好助力可持续发展目标建设

Guo, N., & Ren, Z. R. (2022). Research on the cultivation mode of foreign legal talents under the background of climate change. 黑龙江教育(高教研究与评估), *1387*(5): 71–76. //郭楠,任卓冉.气候变化背景下涉外法治人才培养模式研究.黑龙江教育(高教研究与评估), 2022, 1387(05): 71–76.

Information Office of State Council. (2014, November 25). National climate change plan (2014–2020). http://www.scio.gov.cn/xwfbh/xwbfbh/wqfbh/2014/20141125/xgzc32142/Document/1387125/1387125_1.htm

International Renewable Energy Agency (IREA). (2021). Renewable energy and jobs – Annual review 2021. Technical Report.

Jia, B., He, X. J., & Ge, H. Q. (2011, November 1). Nanxin University to create climate change and public policy research think tank. 中国气象报, p. 003. //贾冰,何晓进,戈华清. 南信大打造气候变化与公共政策研究智库.中国气象报, 2011, November 1, p. 003.

Jiang, L. (2019). Youth climate change action network. *世界环境, 176*(01): 90–91. //姜利.青年应对气候变化行动网络.世界环境, 2019, 176(01): 90–91.

Jiang, Z. H., Weng, D. M., Tu, Q. P., & Others. (2020). Review of research progress on climate and climate change in Nanjing University of Information Science and Technology. 大气科学学报, *43*(5): 785–801. //江志红,翁笃鸣,屠其璞等.南京信息工程大学气候与气候变化研究进展回顾.大气科学学报, 2020, 43(05): 785–801.

Lanzhou University (LZU). (2021). Two professors from Lanzhou University have been named among the world's 1,000 most influential scientists in the field of climate change research. Retrieved August 30, 2024, from http://lzwb.lzbs.com.cn/content/202105/20/content_57943.html//兰州大学.气候变化研究领域全球最具影响力的1000位科学家出炉，兰州大学两位教授入选

Lanzhou University (LZU). (2023). Introduction to the ministry of education's key laboratory on semi-arid climate change. https://atmos.lzu.edu.cn/keylab/lzupage/B20090504091458.html//兰州大学. 半干旱气候变化教育部重点实验室.实验室简介

Li, B. Q. (2022). Research on the cultivation of top innovative talents in universities with industry characteristics – Based on the exploration of Nanjing University of Information Science and Technology. 江苏高教, *254*(4): 52–56. //李北群.行业特色高校拔尖创新人才培养研究——基于南京信息工程大学的探索.江苏高教, 2022, 254(04): 52–56.

Li, P. H. (2016). The theme of "Youth in Action on Climate Change" will be held at the China Corner of the United Nations Climate Change Conference in Paris. *世界环境, 158*(01): 58–59. //李鹏辉."青少年应对气候变化在行动"主题边会在巴黎联合国气候变化大会中国角举行.世界环境, 2016, 158(01): 58–59.

Li, W. L., & Dong, G. H. (2009). Lanzhou University has become an important force in China's climate change research. 气候变化研究进展, *5*(2): 102. //李万里,董广辉.兰州大学成为我国气候变化研究的一支重要力量.气候变化研究进展, 2009: 5(02): 102.

Liao, X. L., Wang, W. F., Zhu, J. W., et al. (2022). Thinking and practice of teaching "Introduction to Global Climate Change" in English. 教育教学论坛, *556*(5): 74–77. //廖晓琳,王维枫,朱将伟等."全球气候变化概论"全英文教学的思考与实践.教育教学论坛, 2022, 556(05): 74–77.

MOEE (Ministry of Ecology and Environment) et al. (2022). National climate change adaptation strategy 2035. http://www.gov.cn/zhengce/zhengceku/2022-06/14/content_5695555.htm. //生态环境部等部门.国家适应气候变化战略2035

MOEE (Ministry of Ecology and Environment). (2023, October 27). China's climate change policy and action 2023 annual report. Retrieved May 21, 2024, from http://sthjj.xlgl.gov.cn/zwgk2022/zcfg2022/202211/P020221104618852441944.pdf //生态环境部.中国应对气候变化的政策与行动2023年度报告

Nanjing University of Information Science and Technology (NUIST). (2023). Introduction. Retrieved July 20, 2023, from https://www.nuist.edu.cn/909/list.htm //南京信息工程大学.学校简介

Renmin University of China. (2012). The launch ceremony of Climate Communication and Climate Change "Into Communities, into Campuses, into Rural Areas, and into Enterprises" and the founding meeting of the Advisory Committee of the China Climate Communication Project Center. 国际新闻界, *34*(6): 127–128. //气候传播与气候变化"进社区、进校园、进农村、进企业"活动启动仪式暨中国气候传播项目中心顾问委员会成立会议在中国人民大学举行.国际新闻界, 2012, 34(06): 127–128.

Shandong University (SDU). (2018). About the centre for climate change and health research. 山东大学学报(医学版), *56*(8): 2+121.//山东大学. 气候变化与健康研究中心简介.山东大学学报(医学版), 2018, 56(08): 2+121.

Shao, L. L. (2020). Implications of the establishment of the World University Climate Change Alliance for university climate education. 中国高等教育, *649*(10): 63–64. //邵莉莉.世界大学气候变化联盟成立对高校气候教育的启示.中国高等教育, 2020, 649(10): 63–64.

Tang, W. J., & Wang, J. (2024). Social practice and model exploration of climate change education in normal colleges: A case study of Life Science College of Capital Normal University. 环境教育, (1): 47–49.//唐文杰 & 王瑾.师范类院校气候变化教育的社会实践与模式探索——以首都师范大学生命科学学院为例.环境教育, 2024, (01): 47–49.

Tsinghua University (THU). (2006). Asia Development Bank establishes Climate Change Knowledge Hub at Tsinghua University. 节能与环保, *11*: 38. //清华大学. 亚行在清华大学设立气候变化知识枢纽.节能与环保, 2006(11): 38.

Wang, C., Cong, J. H., Wang, K., et al. (2021a). Research on the list of technologies for coping with climate change in China. *中国人口·资源与环境, 31*(03): 1–12. //王灿,丛建辉,王克等.中国应对气候变化技术清单研究.中国人口·资源与环境, 2021, 31(03): 1–12.

Wang, H. C., & Wang, Y. J. (2022). Changes and prospects of science and technology policy for adaptation to climate change in China. *Scientific and Technological Progress and Countermeasures, 39*(14): 122–131. //王华春,王英杰.中国适应气候变化的科技政策变迁与展望.科技进步与对策, 2022, *39*(14): 122–131.

Wang, Y. X., Xu, Z. H., & Qi, R. (2021b). On the framework of carbon neutral planning in universities. *中国地质大学学报(社会科学版), 21*(06): 1–9. //王焰新,徐绍红,齐睿.论高校碳中和规划的架构.中国地质大学学报(社会科学版), 2021,21(06):1–9.https://doi.org/10.16493/j.cnki.42-1627/c.2021.06.001

Wu, L. B., & Tian, Y. T. (2020). The innovative development path of high-level industrial characteristic universities under the "double first-class" strategy. 法学教育研究, *30*(3): 347–360. //吴立保,田宇婷."双一流"战略下高水平行业特色高校的创新发展路径.法学教育研究, 2020, 30(03): 347–360.

Xi, J. P. (2020, December 12). Speech at the climate ambition summit. http://www.xinhuanet.com/2020-12/12/c_1126853600.htm //习近平. 在气候雄心峰会上的讲话

Xiao, X. (2012). The international conference on drought disaster monitoring, assessment and management under global climate change was held in Beijing Normal University. 北京师范大学学报(自然科学版), *48*(3): 301. //肖希.全球气候变化下的干旱灾害监测、评估和管理国际会议在北京师范大学召开.北京师范大学学报(自然科学版), 2012, 48(03): 301.

Xue, B. (2014). The Canada-China symposium on urban climate change was held at Northwestern University. 西北大学学报(自然科学版), *44*(4): 520. //薛鲍.中加城市气候变化研讨会在西北大学召开.西北大学学报(自然科学版), 2014, 44(04): 520. https://doi.org/10.16152/j.cnki.xdxbzr.2014.04.045

Yang, K. (2009). An analysis of the actions and impacts of American universities to address the challenge of climate change. 大学(学术版), (12): 77–83+76. //杨珂.美国大学应对气候变化挑战的行动及影响分析.大学(学术版), 2009(12): 77–83+76.

Yang, Y. L., Su, Q., & Yang, R. W. (2022). Greater Mekong subregion climate change research provincial innovation team, Yunnan University. 云南科技管理, *35*(3): 76–77. //杨亚力,苏秦,杨若文.云南大学大湄公河次区域气候变化研究省创新团队.云南科技管理, 2022, 35(03): 76–77.

Zhang, J., & Fan, K. X. (2017). "Design for a Better World" – The first International Symposium on "Design Strategies for Climate Change" was held at Hunan University. 生态经济, *33*(7): 236. //张军,范可馨."为更好的世界而设计"——首届"应对气候变化的设计策略"国际研讨会在湖南大学召开.生态经济, 2017, 33(07): 236.

Zhang, N. X., & Sun, J. H. (2021). Focusing on the construction of "first-class disciplines" to build local industry characteristic university brand – Interview with Li Beiqun, President of Nanjing University of Information Science and Technology. *世界教育信息, 34*(10): 24–26, 35. //张男星,孙继红.聚焦"一流学科"建设 打造地方行业特色高校品牌——访南京信息工程大学校长李北群.世界教育信息, 2021, 34(10): 24–26+35.

Zhang, Z. G. (2017). Witness youth power: The 9th international youth energy and climate change summit was held at Tianjin University. *绿色中国, 481*(15): 72–73. //张志国.拥抱绿色革命,见证青年力量——第九届国际青年能源与气候变化峰会在天津大学举办.绿色中国, 2017, 481(15): 72–73.

Zheng, B. W., & Zheng, Q. (2023). Let climate communication truly form a "big climate" in China and the world: Interview with Professor Zheng Baowei, Director of the China Climate Communication Project Center. *传媒观察, 472*(4): 50–58. //郑保卫,郑权.让气候传播在中国和世界真正形成"大气候"——访中国气候传播项目中心主任郑保卫教授.传媒观察, 2023, 472(04): 50–58.

Zhou, D. C., & Zhang, L. X. (2016). The urgency and necessity of strengthening climate change education in colleges and universities. *Science & Technology Information, 14*(28): 146–147+149. //周德成,张良侠.浅析加强高校气候变化教育的紧迫性与必要性.科技资讯, 2016, 14(28): 146–147+149. https://doi.org/10.16661/j.cnki.1672-3791.2016.28.146

Zhu, S. L., & Zhu, L. (2020). Comments on China's policies and actions on climate change since the 12th Five-Year Plan. *中国人口·资源与环境, 30*(04): 1–8. //朱松丽,朱磊,赵小凡等."十二五"以来中国应对气候变化政策和行动评述.中国人口·资源与环境, 2020, 30(04): 1–8.

CHAPTER 9

The Role of Intra-national Research Collaboration and Funding in Generating Social and Policy Impact on the Amazon Biome

Justin Hugo Axel-Berg, Elizabeth Balbachevsky, and Jacques Marcovitch

Introduction

This chapter focuses on the policy instruments that support changes in the research agenda and design needed to address some of the most persistent issues related to climate change. In fact, redirecting research agendas to address the core issues related to climate change is a central contribution from the higher education sector for combatting climate change. When

J. H. Axel-Berg (✉) • J. Marcovitch
Faculty of Economics and Administration, University of São Paulo, São Paulo, Brazil
e-mail: justin.axelberg@usp.br

E. Balbachevsky
Department of Political Science, University of São Paulo, São Paulo, Brazil
e-mail: balbasky@usp.br

J. Kekäle, R. Pinheiro (eds.), *Higher Education Policy for Tackling Climate Change*, https://doi.org/10.1007/978-3-031-78479-8_9

analysing the Brazilian experience, this chapter argues that the needed changes are not only related to the research agenda but, critically, to reshaping the patterns of collaboration linking more local research networks and the more internationalised ones. As we argue at the chapter's end, to advance new approaches sustaining a better balance for the participants from local research networks, funding agencies need to develop fine-tuned policy instruments that could nurture and support local actorhood, allowing for the local research agenda to be prioritised in these research efforts.

While Brazil ranks as the fifth largest emitter of greenhouse gases globally, it is also one of the most important parts of any potential global solution to the crisis. Therefore, the urgency to address this issue is paramount. A significant portion of our emissions (75%) stem from land use and cover changes, particularly the deforestation of vast areas in the Amazon biome (Azevedo, 2024). To meet our commitments under the Paris Accords, it is imperative to establish a new bioeconomy that values the standing forest in the Amazon biome (Gatti et al., 2021).

Limiting deforestation requires the development of production chains that can add value to sustainably produced non-timber forestry products. This will allow the development of an inclusive economy that brings prosperity to the region while guaranteeing environmental protection and regeneration. One condition is deemed necessary to achieve these goals: the development of a cluster of knowledge-intensive organisations anchored in the region's local knowledge that could guide the sustainable exploration of the region's potential.

The potential benefits of sustainable bioeconomy activities are vast. They are knowledge-intensive, requiring a fusion of scientific research, public and private sector development, local knowledge, and experience. A productive and forest-protective chain must also provide robust public healthcare (Castro, 2022), educational opportunities and training for communities, and social welfare. These elements are all correlated with increased economic growth and reduced deforestation rates, promising a brighter future for the Amazon biome.

Despite this, institutions in the Brazilian Amazon biome produce only around 2% of Brazil's total articles and just 3% of its PhDs. The universities inside the Amazon biome hold a vital position for their ability to act as an interface between scientific research and local knowledge. Their proximity to the forest should mean they could explore topics closely aligned with local priorities and small producers rather than institutions outside the biome. Ciarli and Ràfols (2019) found similar results comparing research priorities in research on rice.

This chapter follows the recent developments of one regional production chain—the *Arapaima gigas* (pirarucu) fish chain—that experienced considerable changes in the direction of a more sustainable approach, focusing on the critical roles played by several local knowledge-intensive institutions in guiding these changes while, at the same time, mapping some frailties in these dynamics that should be addressed by well-crafted policies. To achieve these goals, this article adopts the following structure: first, it maps the main features of the Brazilian science, technology, and innovation (ST&I) ecosystem, specifying the main characteristics of the Amazonian ST&I ecosystem and exploring the role played by local and national research institutions in developing innovative approaches to the traditional pirarucu's value chain. By evaluating the roles played by different actors in changing this traditional value chain, the chapter explores the achievements and shortcomings of the contemporary policies for developing a sustainable bioeconomy in the Amazon biome and the challenges this change poses for higher education and research policies.

Brazil's Science, Technology, and Innovation Ecosystem: An Overview

The public sector dominates the Brazilian science, technology, and innovation ecosystem. The most critical sites for knowledge production inside the country are, by far, the academic environments that gravitate around the postgraduate system inside the public universities. As in most Latin American Universities, postgraduate education is a late addition to the regional university institutional fabric. The traditional university in Latin America followed the Napoleonic heritage; it was organised around professional bachelor programmes whose diplomas were, at the same time, an academic degree and a professional certification, usually granting the holder access to protected areas in the labour market. It was only in the 1960s that this picture experienced relevant changes in many countries, with the establishment of national councils to support research. While the Brazilian experience follows these general patterns, there is a notable difference: in the Brazilian case, research entered the university fabric at the same time as postgraduate education—master's and doctoral degrees—was recognised and regulated by law. In the Brazilian experience, postgraduate programmes became the privileged locus for institutionalising research in Brazilian higher education (Balbachevsky, 2013).

At the beginning of the 1960s, the Ministry of Education recognised the new degrees and established a dedicated structure to support graduate education. At first, through a Campaign designed to support the academic qualification of the faculty employed at public universities, CAPES—the Campaign for Qualification of the Higher Education Personnel—quickly evolved into a dedicated agency providing scholarships, supporting, and evaluating the newly created graduate programmes. In the same period, the federal government directed considerable resources to research as part of the strategies for pushing the country's development. A relevant part of these resources ended up supporting research inside the high-ranking postgraduate programmes, mainly because, at the time, these programmes provided a suitable environment for attracting and retaining Brazilian scholars coming from abroad after finishing their doctoral training.

These historical dynamics were responsible for introducing an informal but highly visible stratification among Brazilian public universities: at the top, a small number of universities, some owned by the Federal government, others owned by state governments, experienced early differentiation, securing scholars with good academic credentials, developing a large and dynamic postgraduate tier and offering good research infrastructure. Because of these conditions, these universities attracted significant research funds from federal and state-level agencies for research support. Owing to the well-known mechanism of the "Matthew effect" first described by Robert Merton (Merton, 1968), these universities became Brazil's most important sites for research. Not surprisingly, these institutions are concentrated in the south-east region, the wealthiest region in Brazil. Even in 2022, the four states in this region accounted for 52% of all doctoral degrees granted in the country. Inside this region, the State of São Paulo granted more than 27% of all doctoral degrees in the country (GEOCAPES, n.d.).

Brazilian S&T Public Policies: A Layered, Fragmented System

The first steps in building a system supporting scientific research in Brazil date from the 1950s. The Brazilian National Research Council (CNPq, a Portuguese acronym for Conselho Nacional de Pesquisas) was funded in

1951; in that same year, the Federal government also launched the National Campaign for the Development of Higher Education Personnel (CAPES), which in time became the most relevant agency supporting, evaluating, and framing graduate education in Brazil. Between the 1950s and the beginning of the 1970s, this system remained small, underfunded, and primarily focused on supporting a few small-scale projects, mainly in physics, with a narrow focus on atomic energy.

The most significant leap forward came in the 1970s when the Brazilian most important investment Bank, the BNDES (Portuguese acronym for *Banco Nacional de Desenvolvimento Econômico e Social*—National Bank for supporting economic and social development) involved itself with policies for funding research and science, launching a dedicated fund with substantial resources targeting specific projects deemed necessary for the country's development. The Fund, an expanded version of CNPq, and a new agency—FINEP—were put under the Ministry of Economic Planning, creating, for the first time, a continuous flow of resources supporting laboratories, postgraduate programmes, and other initiatives. This support gradually created a stable institutional space for research and knowledge production in some public universities.[1]

A lasting feature coming from these first attempts to institutionalise public research support is the adoption of project grants and scholarships as the sole instruments for supporting research activities inside universities. A research grant may include several undergraduate science initiation scholarships, master's, doctoral, and post-doc scholarships, and even some technical scholarships that allow the research group to enlist the help of technicians, but it never allows for work contracts. It can also grant resources for acquiring equipment and materials, IT infrastructure, etc. However, all budgeting and spending are linked to particular research projects, all with a beginning and an end. They are always limited in time and are supposed to produce some deliverables, primarily published material.

In the 1990s, all federal agencies supporting research came under the umbrella of the newly established Ministry of Science, Technology and

[1] At the end of the 1960s and early 1970s, most academics in Brazil did not hold doctoral degrees. Thus, restricting access to these resources to researchers holding a doctoral degree limited access to research support to a few public universities (federal and state-owned) and some denominational ones. These universities successfully attracted the first generation of PhD holders returning from their studies abroad and established the first graduate programmes.

Innovation.[2] A new layer of policy instruments was added when these agencies started to launch calls to steer the research agenda towards social, technological, environmental, and/or economic issues. By the end of the 1990s and the 2000s, these agencies also adopted calls for supporting long-term research projects and regional/or local research networks. Despite these changes, Brazil's science policy's central feature remained: the direct support to the researcher through projects and the minor involvement of the universities in supporting research activity.

The second feature of Brazilian research institutional building is the great degree of autonomy enjoyed by research sites inside the universities. In a higher education system primarily designed for teaching and revolving around training at the bachelor level, research established itself as small semi-autonomous and partially institutionalised spaces—laboratories, research groups, centres, etc.—using a parallel budget which came directly from the funding agencies to the projects' heads, bypassing the university budgetary process; and with rules and a prestige ladder that was only partially connected with the academic career recognised by the university. Graduate education developed attached to these research units and enjoyed a similar situation of semi-independence, supported by resources and scholarships coming directly from CAPES and from other agencies for research funding, also bypassing the university's budget (Balbachevsky, 2000; Durham & Gusso, 1991; Schwartzman, 1991).

Together, these two dynamics helped create a peculiar institutional logic (Thornton et al., 2012) linked to the Brazilian research community: a highly entrepreneurial and strongly independent ethos, valuing the autonomy of their small research unit above all. This autonomy is a core value for the Brazilian scientific community. They are willing to fight for this autonomy even against the macro-institutional environment in which their research unity is situated, be it a university unit or the entire university (Balbachevsky, 2008).

The growth and diversification of the public agencies responsible for supporting research described above created new opportunities for the research community to establish themselves as the main interlocutor for the public officials in charge of designing and implementing programmes supporting research and technological development. Moreover, in this

[2] CAPES, the main agency in charge of regulating, evaluating, and supporting graduate education—a key institutional site for research in Brazil, remained under the control of the Ministry of Education.

process, funding agencies imported the main traits of the institutional logic described above to their institutional features. As we will see in the case described below, this dynamic presents no problem when the delegation contract between researchers and agencies follows what Braun (2003) characterised as a blind delegation, where agencies trust science to choose the research agenda and modes of knowledge production. But once these agencies create their output expectations and policy agenda, the situation becomes more conflictive, and the traditional arrangement's limitations start to be visible.

The Federative Counterpart in Research Funding

Brazil is a Federative State. Local and regional dynamics are essential in all policy fields, as well as for research funding policies. Besides the federal layer of agencies designed for supporting research and graduate education, all states in Brazil have their own science foundations.

São Paulo, the wealthiest state of the Federation, launched FAPESP, the State of São Paulo Science Foundation, in 1962. From the beginning, the state's Constitution established that FAPESP would control at least 0.5[3]% of the state's ordinary revenues and that these funds should be used to finance research and technological development. In the following years, other states launched their own Science Foundations, and in 2006, a National Council of State Research Foundations (COFAP) was established to coordinate the policies and actions of all regional science foundations.

The degree of support each foundation receives from its state government varies greatly. Some of these foundations are well structured and firmly institutionalised, with a well-established bureaucracy, stable sources of financial support, and a degree of institutional insulation that allows for long-term policy plans. Others are small, weakly institutionalised, and open to the direct effects of local political battles. Nevertheless, they all operate with a regional focus, linking their legitimacy to advancing and supporting a research agenda targeting local problems.

The federal agencies—CNPq, FINEP, and CAPES—are used to launch calls and programmes in partnership with state science foundations. In fact, in many programmes focused on buster efforts around more

[3] The new state constitution from 1989 raised this percentage to 1% of the state's ordinary revenues.

ambitious research agendas, it is usual that the federal agencies establish consultations with COFAP and/or specific regional foundations when starting to design a new programme. These consultations aim to add resources to the programme and ensure the mobilisation of local regional research teams to answer the call.

Another relevant venue more recently explored by COFAP is the establishment of cooperative agreements and programmes linking research efforts supported by different state science foundations. The main goal of these agreements is to support the development of regional research capabilities, reinforcing cross-region research networks and advancing a research agenda focused on regional issues. The programme this chapter evaluates is one of this sort. The agreement between FAPESP and the Amazonas State Science Support Foundation (Fapeam) aimed to develop and support cooperative scientific and technological research projects linking researchers from the state of São Paulo and the State of Amazonas. It was launched in 2019, and the programme's central goal was to promote and support a research agenda of a new bioeconomy for the Amazon biome (FAPESP, 2020). One of the most relevant features of the programme was its aim to connect research teams from the more research-intensive universities in Brazil (the public universities in São Paulo) with researchers from the Amazon institutions that work in a much less internationalised environment but, on the other hand, have a deep understanding of the Amazonian region. Our chapter explores the impacts of this collaboration in reshaping the regional research agenda on a particular—but central—theme for the sustainable development of the Amazonian region, which is fishing.

Scientific Production in the Brazilian Amazon Biome

The legally defined Brazilian Amazon region is composed of nine states—Acre, Amapá, Amazonas, Maranhão, Mato Grosso, Pará, Rondônia, Roraima, and Tocantins. However, because only small parts of Mato Grosso reside within the biome, and these parts do not contain research institutions, we have not considered them in this analysis. As described in Table 9.1, there are 12 federal universities within the biome, five state universities, some significant research institutes, both federal and private, and one research-active private university. According to CAPES, these institutions had 114 PhD granting programmes, which granted 859

Table 9.1 Publications by research institutions in the Amazon 2018–2023 Dimensions database

Name	*Publications*	*Citations (mean)*	*Institutional status*
Federal University of Para (UFPA)	23,310	10.56	Federal
Federal University of Amazonas (UFAM)	9020	9.77	Federal
Federal University of Maranhão (UFMA)	8627	8.68	Federal
National Institute of Amazonian Research (INPA)	7791	23.6	Institute
Federal University of Tocantins (UFT)	4734	5.43	Federal
Federal Rural University of Amazonia (UFRA)	2587	7.37	Federal
State University of Pará (UEPA)	2563	6.91	State
Instituto Evandro Chagas (IEC)	2507	22.91	Institute
Museu Paraense Emílio Goeldi	2408	24.77	Institute
Federal University of Rondônia (UNIR)	2406	7.61	Federal
Universidade Federal do Acre (UFAC)	2314	13.21	Federal
Federal University of Amapá (UNIFAP)	1955	7.5	Federal
Federal University of Western Pará (UFOPA)	1842	9.05	Federal
Federal University of Roraima (UFRR)	1658	6.63	Federal
University of the State of Amazonas (UEA)	1137	12.05	State
Federal University of Southern and Southeastern Pará (UNIFESSPA)	1047	3.94	Federal
Universidade Estadual do Maranhão (UEMA)	603	7.4	State
Centro Universitário do Maranhão (UNICEUMA)	439	8.81	Private
Universidade Nilton Lins	408	11.86	Private
Instituto Federal do Amazonas (IFAM)	405	5.22	Institute
Federal Institute of Pará (IFPA)	381	5.87	Institute
Universidade Estadual do Amapá (UEAP)	327	4.69	State
Universidade Estadual de Roraima (UERR)	296	3.75	State
Instituto de Desenvolvimento Sustentável Mamirauá (IDSM)	293	15.18	Institute
Instituto de Pesquisa Ambiental da Amazônia (IPAM)	263	95.95	Institute

Source: Dimensions database, 2023, processed by the authors

doctoral degrees in 2022, representing 4.6% of all PhD granting programmes and 3.7% of all doctoral degrees granted in the country that year.

This table shows that two flagship federal universities predominantly drive the research output, the Federal University of Pará (UFPA) and the Federal University of Amazonas (UFAM), with a set of three research institutes supplementing this strength: the National Institute of Amazon Research (INPA), the Evandro Chagas Institute, and the Emilio Goeldi Museum. Considering bioeconomy-relevant research, the role of Mamirauá (IDSM) and the Institute of Amazon Environmental Research (IPAM) are also noteworthy. While the two universities are responsible for the majority of the output, the research institutes produce what tends to be the most highly cited research in the biome.

Methodology

To map the contribution of the research institution situated in the Amazon biome to the bioeconomy-relevant literature related to the Amazon region, we used the Dimensions database (Digital Science, n.d.).[4] This database was selected for its much larger publication spectrum than Scopus or Web of Science and better representation of regional sources, including SciElo, conference proceedings, and pre-print archives.

An initial protocol was designed using the keywords:

((amazon) AND ((economy OR bioeconomy) OR (("sustainable development") OR (development) OR (innovation))*

It was then manually refined by removing keywords that produced erroneous results:

*NOT dam NOT neurosurgery NOT alcohol NOT indoor, NOT electrocardiography, NOT chagas, NOT COVID, NOT ux, NOT snake, NOT Alzheimer**

And fields of research unrelated to our area of interest:

NOT (21 History and Archaeology OR 22 Philosophy and Religious Studies OR 1103 Clinical Sciences OR 0803 Computer Software OR 0206 Quantum

[4] Dimensions is a comprehensive database containing abstracts, citations, and research grants linked to corresponding publications, clinical trials, and patents. Produced by Digital Science, a technology company based in London (UK).

Physics OR 0103 Numerical and Computational Mathematics OR 0307 Theoretical and Computational Chemistry OR 0802 Computation Theory and Mathematics OR 19 Studies in Creative Arts and Writing OR 13 Education)

As it is possible to see in Table 9.2, despite its relative weakness both in terms of the number of active research institutions and the number of active research sites, using the number of doctoral programmes as a proxy, the research performance of the research organisations situated in the region of the Amazon Biome is impressive when one isolates the topics relevant for the Amazonian bioeconomy. In fact, in the past five years, 26.1% of all research on Amazon bioeconomy published by Brazilian research organisations had the participation of an institution situated in the Amazon biome. This increased from 22.3% in 2007–2011 to 19.7% in 2012–2016. These figures show a gradual but constant change in the same direction, asserting the growing participation of the Amazonian research community in exploring critical issues related to the region's sustainable development.

Although this seems modest in relative terms, it should be noted that the absolute expansion has been massive. The overall dataset grew by 75% between the first five-year period and the second, and 54% in the second. This growth rate is far above the global average for increasing publication.

These data provide some evidence of a healthy pattern of specialisation of the Amazonian research institutions in topics relevant to the need for

Table 9.2 Articles in Brazilian bioeconomy 2007–2021

Period	*Published by an Amazonian institution*	*Number of articles*	*% increase from the previous period*	*% of the total for the period*
2018–2022	Yes	11,637	80.1	26.1
2018–2022	No	32,925	46.1	73.9
2018–2022	Total	44,564	53.7	
2013–2017	Yes	6463	98.6	22.3
2013–2017	No	22,533	69.7	77.7
2013–2017	Total	28,996	75.3	
2008–2012	Yes	3255		19.7
2008–2012	No	13,282		80.3
2008–2012	Total	16,537		

Source: Dimensions database, processing done by the authors

developing a sustainable bioeconomy in the region. In fact, despite their small number, this research community has an impressive and growing impact on the knowledge production in this area inside the Brazilian community. However, some drawbacks are important to note. First, considering the knowledge production chain that goes from the basic task of collecting data and setting up databases to the high-end of securing a new patent, what are the most usual positions occupied by the researchers from organisations situated in the Amazonian biome? Again, the data collected by the Dimensions database offers some clues that allow us to answer this question. As one can see in Table 9.3, since 2008, at least, researchers working at research organisations situated in the Amazonian biome are responsible for publishing more than 50% of the bioeconomy-related open datasets produced in Brazil. This participation declined from 70.8% to 57% for the entire period from 2008 to 2022. However, this research community is present in producing critical information for developing a new bioeconomy in the Amazon region.

Notwithstanding, this research community plays a secondary role at the high-end of the knowledge producing chain. As Table 9.4 shows, between 2007 and 2021 the number of patents produced in Brazil related to the bioeconomy grew from a total of 202 patents granted in the period of 2007–2011 up to a total of 628 patents granted in the period of 2017–2021, an impressive growth of 310%. During this period, considering all patents related to bioeconomy granted in Brazil, the participation of the Amazonian institutions grew from 3% to 8.8%. An important

Table 9.3 Bioeconomy-related open datasets produced in Brazil 2008–2022

Period	*Published by Amazonian institutions*	*Datasets*	*% of previous period*	*% of the total for current period*
2018–2022	Yes	813	209.1	57.0
2018–2022	No	613	308.7	43.0
2018–2022	Total	1426	245.3	
2013–2017	Yes	263	1447.1	63.7
2013–2017	No	150	2042.9	36.3
2013–2017	Total	413	1620.8	
2003–2012	Yes	17		70.8
2008–2012	No	7		29.2
2008–2012	Total	24		

Source: Dimensions database, processing done by the authors

Table 9.4 Bioeconomy-related patents produced in Brazil 2007–2021

Period	*Published by Amazonian institutions*	*Number of patents granted*	*% increase from the previous period*	*% of the total for current period*
2017–2021	Yes	55	120.0	8.8
2017–2021	No	573	51.6	91.2
2017–2021	Total	628	55.8	
2012–2016	Yes	25	316.7	6.2
2012–2016	No	378	92.9	93.8
2012–2016	Total	403	99.5	
2007–2011	Yes	6		3.0
2007–2011	No	196		97.0
2007–2011	Total	202		

Source: Dimensions database, processing done by the authors

increase, but one which still leaves the research community in the Amazonian region almost excluded from the profits produced at the high end of the knowledge production chain related to the bioeconomy.

In summary, Amazon institutions play a relevant role in contributing to the published material derived from the knowledge related to the new bioeconomy in the region, and they also produce a large proportion of the open datasets compared to Brazilian institutions as a whole—this suggests that they are much more engaged in primary research and data collection than other institutions. This offering of open datasets forms an important resource for other scientists outside of the biome lacking access to the forest and its populations.

If Amazonian institutions carry out a large duty of collecting primary information, they do not produce as many patents as institutions in other regions. This means that many relevant technological solutions are being produced outside the region, which poses a particular policy issue in the effort to create a sound knowledge base for sustaining the development of a new bioeconomy, a key question in Brazil's strategy for fighting climate change.

The Case of Pirarucu

Among the activities involved in bioeconomy, fishing is one of the most significant, for its role in maintaining traditional community practices, protecting the standing forest, supplying a significant source of food for local communities, and significant potential for commercialisation. Among the traditionally exploited species in the Amazon biome, the *Arapaima Gigas*, or pirarucu, has one of the highest potentials for insertion into a sustainable bioeconomy. The pirarucu is one of the world's largest fresh-water fish, reaching up to three metres in length and over 300 kg in weight (Castello & Stewart, 2010; Imbiriba, 2001). As well as playing a fundamental role in the subsistence of local communities and preserving biodiversity, the fish has significant market potential as a source of protein for the local community and as a high-value food for export. Furthermore, the processing procedure generates valuable by-products such as consumer leather and fishmeal for animal feeds and fertiliser. Therefore, it offers the potential for high-quality employment in conservation-related activities and depends on preserving the standing forest around the river, and clean waterways for its habitat.

By the mid-1970s, the pirarucu had become critically endangered due to overfishing, pollution, and habitat loss. In response to this situation, legal, commercial exploitation of the species was prohibited and has been gradually reintroduced over the following decades. Today, with stocks gradually recovering, there are two main methods of producing pirarucu for commercialisation: managed fishing on determined reserves and aquaculture (fish farming) (Saes, 2024).

We selected the pirarucu as a chain for analysis because both activities require a combination of scientific knowledge to be developed—especially on reproductive behaviour, sexing, feeding, rearing, monitoring stocks, setting fishing quotas, slaughter, processing, processing of by-products, preservation and refrigeration, among others. Alongside this need for scientific knowledge, traditional knowledge plays an important role in the production chain of the pirarucu, which has made up a significant part of local practices and diets for millennia (Saes, 2024).

We produced a search with the following protocol in the Dimensions database—chosen for its greater coverage of local resources, including SciELO. It should be noted, however, that institutional repositories are also significant sources of publication in the Amazon biome. They tend not to be formally indexed, and do not have DOIs attached to contents, so were not included in this study.

A search was carried out with the following terms:

(pirarucu OR arapaima OR arapaima gigas OR paiche) AND (production OR cultivation OR industrialisation OR processing OR economy OR bioeconomy OR management) AND (amazon)*

While there are significant quantitative differences in the behaviour of institutions inside the Amazon biome and those outside it, this section looks at whether there are qualitative differences between how universities inside the biome produce knowledge when publishing alone (or in coauthorship with another institution inside the biome); when publishing with a coauthor from an institution outside the biome; and Brazilian institutions publishing research on a topic with high relevance to bioeconomy, but without a coauthor in the biome. For this exploratory research, the publications on a single production chain were chosen because it allows us to view different approaches to similar topics.

Full records were extracted from the Dimensions database, cleaned for relevance, and visualised using the VOSViewer, v1.6.19 (Van Eck & Waltman, 2007). The algorithm clusters key terms together, with the node's size indicating the term's frequency, and the edge indicating where the terms appeared together in the same abstract. This gives us a conceptual map of a set of articles (Table 9.5).

The Dimensions database contains 864 articles about managing pirarucu from the biome. When all articles with coauthorship from outside the biome are filtered out of this dataset, we are left with 300 publications (Fig. 9.1).

This topic map shows that when the biome carries out research on its own, the contents are split roughly evenly into the two main forms of cultivation of the fish—the red cluster applies to techniques in sustainable management of wild fishing, while the green cluster refers to techniques in aquaculture and farming. The absence of scientific vocabulary reinforces the fact that this knowledge is highly applied, of low technological

Table 9.5 Publications on Pirarucu on Dimensions database, 2018–2023

Location of authors	*Publications*
Inside the Amazon biome, no outside collaboration	300
Inside Amazon, with outside collaboration	864
Outside the Amazon biome, no collaboration inside	1181

Source: Dimensions database, 2023, processed by the authors

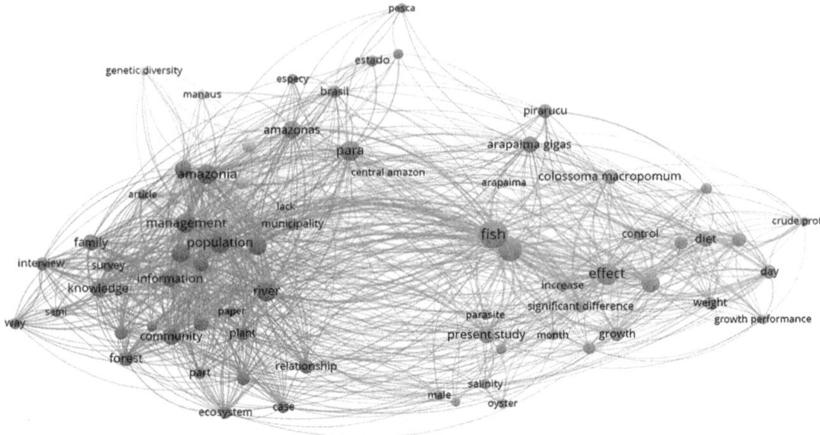

Fig. 9.1 Keyword map of publications on pirarucu from institutions inside the Amazon biome

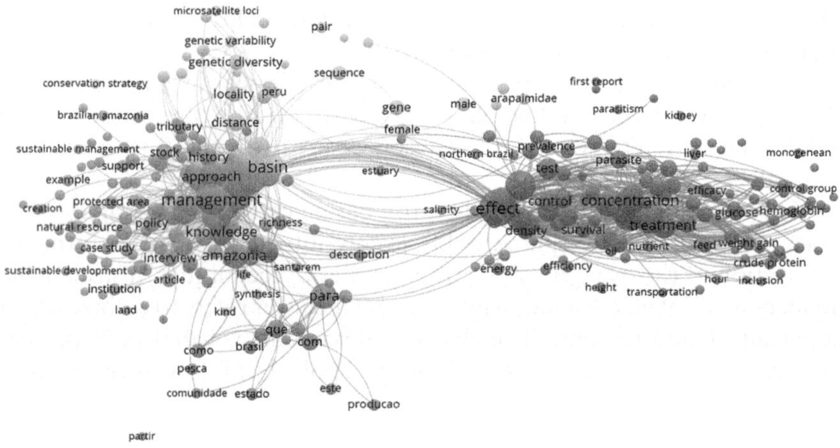

Fig. 9.2 Keyword map of publications on pirarucu from institutions inside the Amazon biome collaborating with institutions outside the biome

burden, and geared towards direct application. This shows that when working alone, institutions in the Amazon have a tendency to produce knowledge that is low complexity and suitable for immediate application (Fig. 9.2).

This figure, which represents a relational topic map of all 864 articles from the Amazon biome shows that the research priorities are significantly different. While there are still two clusters, one each dedicated to aquaculture and managed sustainable wild fishing, we see a much higher prominence for scientific terminology related to the physiology of the pirarucu, represented by the presence of keywords such as "liver", "kidney", and "parasite". The cluster related to sustainable wild fishing, for its part, has a much lower prevalence of words related to individual communities such as "family", "community", "forest", and "relationship", and a much greater role for themes related to policy and management, such as "sustainable development", "protected area", "institution", and "sustainable management".

In the yellow cluster, we also see keywords such as "genetic variability", "genetic diversity", "sequence", and "gene". The words "microsatellite tool" also appear in this cluster. This shows an interest in genetics and breeding, as well as highly technologically sophisticated methods of tracking and monitoring the fish (Fig. 9.3).

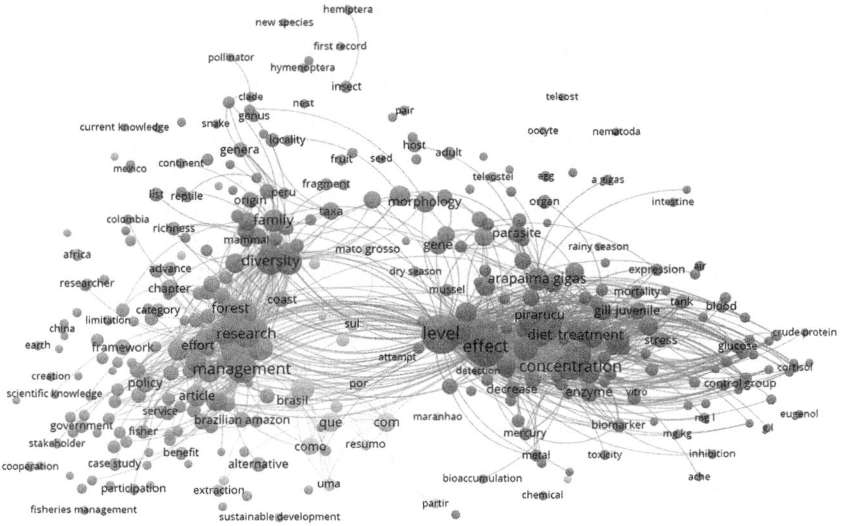

Fig. 9.3 Keyword map of publications on pirarucu from institutions outside the Amazon biome with no collaboration with institutions inside the biome. (Source: Dimensions database, 2023, processed by the authors)

Finally, this topic cluster map is made from the 1181 articles about pirarucu published in Brazil, but without the participation of any institution in the Amazon. The topics that these articles approach are dramatically different from the first figure and markedly different to the second. For the green cluster concerned with environmental management, there are no references to family, community, or fishing, but keywords such as "scientific knowledge", "framework", "category", and "government" predominate. In the red cluster dedicated to the physiology of the individual fish, relevant to those engaged in aquaculture, we see a much lower prevalence of keywords such as "weight gain", "diet", "parasite", and "survival" and a much greater prevalence of words related to biological sciences such as "glucose", "enzyme", and "biomarker". Equally significant is that this is the only figure to feature the words "mercury", "metal", and "bioaccumulation"—one of the most significant risks to the survival of the waterways in the region due to illegal gold mining. Finally, the cluster related to genetic diversity is much larger in this figure than in the second. This shows that significantly more laboratory work is published outside the biome than inside it.

The production system of relevant scientific knowledge on the pirarucu in Brazil demonstrates distinct topics and research agendas emerging from each of the three types of production: institutions in the Amazon working alone, institutions in the Amazon working with institutions in Brazil, and institutions in Brazil working without the input of institutions in the biome.

Institutions in the Amazon not collaborating with institutions from outside of it broadly produce knowledge that is relevant to the immediate needs of producers engaged in aquaculture and sustainable fishing activities. It is also the only institutional arrangement that is significantly concerned with the role of traditional societies, and the role of the family in small production. However, this research tends not to require a high level of scientific intensity in terms of laboratory equipment and financial resources; neither is it specifically concerned with informing policymaking. The research capabilities observed in the Amazon are not sufficient to meet the knowledge demands of a sustainable and profitable production chain of pirarucu.

Institutions in the Amazon play a vital role in the application of knowledge that can empower traditional communities to improve their well-being, pursuing traditional activities and leaving the forest standing and rivers running clean, but they do not fulfil the scale of demand for new scientific knowledge on their own.

When research organisations in Brazil produce research on the pirarucu without an organisation's participation in the Amazon, they tend to pursue topics related to biological sciences, the physiology of the pirarucu, genetics and breeding, and macro-level policymaking. However, this high level of technological and scientific intensiveness tends not to contain clear signs that the local producer communities are considered, and less mention of direct and applicable results. Therefore, it suggests that research organisations outside the biome will find applying the results of their research more challenging than organisations inside of it.

In the second instance, we see a more balanced set of research interests and priorities that combine scientific intensiveness, concern with physiology, biology, and genetics of the fish that are vital to make a sustainable production chain viable, with practical and applied knowledge that can be utilised by the traditional populations who produce the fish. Likewise, it combines a concern with policymaking, use of technology for monitoring stocks and numbers combined with a greater research interest at the local level.

As Fig. 9.4 makes clear, there are a number of distinct clusters for the production of knowledge on the pirarucu, with Brazilian clusters to the right of the map, and international coauthors predominantly on the left.

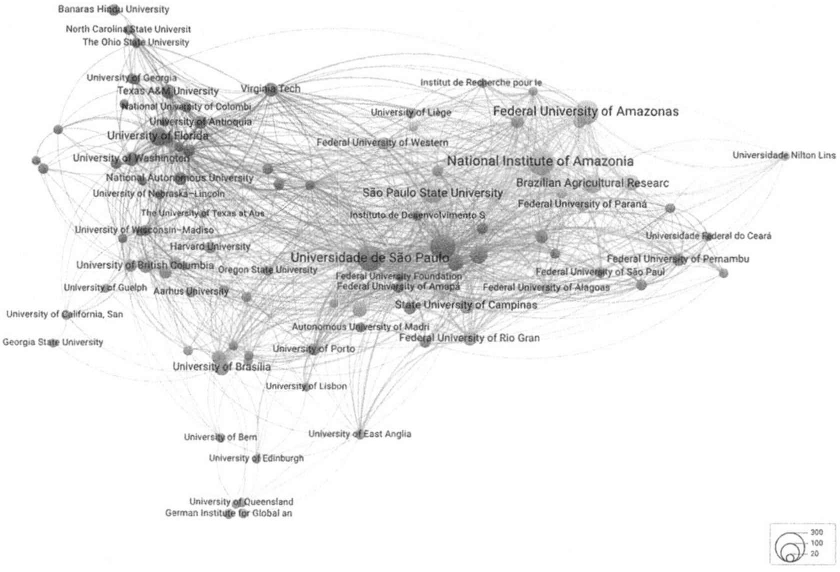

Fig. 9.4 Brazil and the international science network for pirarucu

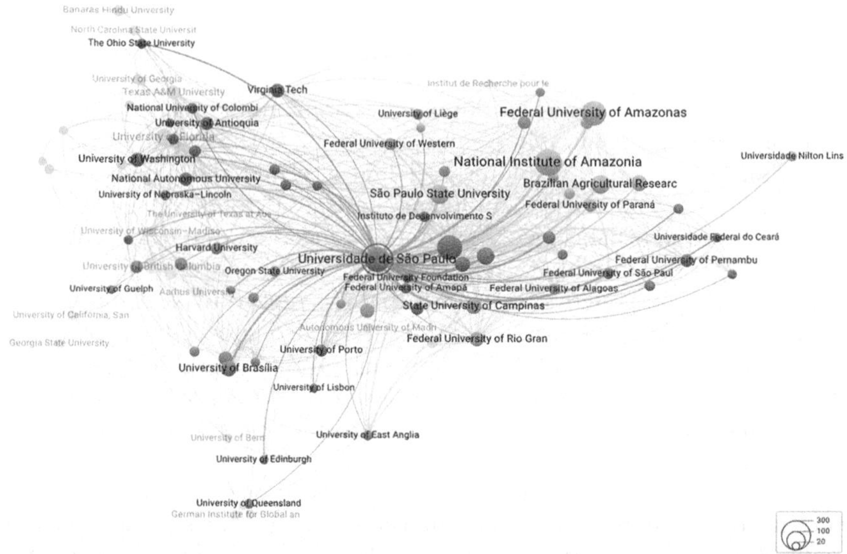

Fig. 9.5 The University of São Paulo's position in global research network on pirarucu

The same map with the University of São Paulo's collaboration relations highlighted, we can see that it acts as a mediating institution between institutions in the biome and those internationally—it collaborates with almost all institutions and serves as a central bridging point between institutions in the biome and international collaborators. Therefore, we can see that just as institutions in the Amazon act as key bridging institutions that connect researchers in the southeast of the country to local stakeholders, allowing greater social impact, institutions in the southeast act as aggregators and gateway institutions for international collaborators looking to work in the biome (Fig. 9.5).

Discussion and Conclusion

The issue of climate change is the most urgent threat facing humanity in the first half of the twenty-first century. Transition away from the use of greenhouse gas-emitting fossil fuels, reducing carbon footprints, providing education to future generations in sustainability, and contributing to the global scientific effort to monitor and mitigate the impact of human activity form a crucial part of universities' activities the world over. As argued at the chapter's beginning, in the case of Brazil, developing the bases for a new bioeconomy in the Amazonian Biome is a central policy issue, since most of our greenhouse emissions come from deforestation and land use change.

As stated in this book's introduction, developing a new research agenda focused on green transition is a key aspect of higher education transition for a new green bioeconomy. This chapter argues that the core needed changes are not only related to changes in the research agenda by itself but mostly a change in the patterns of collaboration linking higher education institutions and researchers based in crucial areas for conservation and the wider global networks of research. We develop this argument by looking at the patterns of collaboration between higher education institutions and researchers located in the Amazonian biome and nation and global research networks, with special attention to the patterns of collaboration between institutions and researchers localised in the biome and those localised in richer and better endowed higher education institutions in Brazil's southeast region.

In Brazil, universities have an obligation of crucial importance—producing and applying the knowledge required to ensure the protection and survival of the Amazon rainforest, one of the earth's largest carbon sinks, its greatest source of biodiversity, and Brazil's single largest source of carbon emissions due to land use change from deforestation. In order to protect the forest, its 20 million inhabitants need sustainable jobs that generate value from the standing forest and clean waterways; construction of these sustainable jobs in the bioeconomy requires high levels of qualified labour and, crucially, the production of scientific knowledge that can be used to improve the functioning of production chains and inform the regulatory policy that governs them.

This chapter looks at the different roles taken by elite traditional research-intensive institutions in southeast Brazil (particularly in the metropolises in São Paulo and Rio de Janeiro) and research organisations in the Amazon region. Due to a historical imbalance of funding and capacity building, higher education institutions in the Amazon region, which has some of the highest demands for science on earth, have been underfunded and underdeveloped. While this scenario is rapidly changing, intranational collaboration has proven to be a vital mechanism for sharing expertise and facilities that these institutions would otherwise struggle to access. Institutions in the Amazon region have witnessed a massive increase in bioeconomy-related research in the last five to ten years, and this has now constituted a significant amount of the knowledge produced on sustainable development of the biome. However, the institutions do not possess the resources to satisfy the demands for scientific knowledge needed to create a sustainable bioeconomy.

There are not only quantitative and distributional differences that this collaborative pattern helps to ameliorate but also distinct qualitative and semantic approaches to bioeconomy that each institution type pursues. In order to explore these issues, this chapter analyses the published outputs of the research networks that work with the *Arapaima Gigas*, or pirarucu, a traditionally exploited species of fish which has one of the highest potentials for insertion into a sustainable bioeconomy.

Research related to sustainable fishing of the pirarucu, when it is produced entirely within the biome, tends to have a technical and vocational angle and a concern with the role of small producers. In contrast, research on the same topic done outside the biome is heavily concerned with physiology, biology, and genetic profiling of fish and policymaking at a national level. However, articles that count on authors coming from inside and outside the biome tend to show a more balanced approach, combining the interests and producing research with the potential for local and national impact. This finding reinforces the need for collaboration between institutions in Amazonia and the wider world.

Finally, a broader debate needs to happen on how scientific credit is distributed for sustainability-related research. Institutions in the Amazon are responsible for 90% of the open datasets that inform the research of others. However, just 9% of patents are derived from research on products from the forest. They are underrepresented as first authors and often appear as junior partners in research projects. This speaks to a scientific tradition that values the production of novel findings over their implementation and laboratory work over fieldwork in the biome. This consequently leads their national and international collaborators to receive a more significant share of the scientific credit for work that would, in all likelihood, not have the same potential for application without the contribution of institutions inside the biome.

For these reasons, we highlight the importance of joint funding initiatives between the São Paulo State Research Council (FAPESP) and the Amazonas State Research Council (FAPEAM) in recent years, in both the FAPESP/FAPEAM joint funding initiative and Amazonia +10. These initiatives recognise the importance of the dynamics identified in this chapter and seek to reinforce them by funding joint research proposals from the State of São Paulo and the State of Amazonas to conduct research in bioeconomy-related subjects (Marcovitch & Val, 2024). However, the chapter also sheds light on the limitations of these initiatives, framed as they are by the usual approach adopted by public research funding in Brazil: projects selected through broad thematic calls, when combined with an institutional environment that insulates leading research units, leaves ample space for strategic choice in the hands of the researchers themselves, and tends to reinforce more traditional approaches to building the research agenda and attributing roles inside the research network. To advance new approaches sustaining a better balance for the participants from the local research networks, the funding agencies need to develop fine-tuned policy instruments that could nurture and support local actorhood, allowing for the local research agenda to be prioritised in these research efforts.

References

Azevedo, T. (2024). "As fronteiras do desmatamento". Interview given to Marcos Pivetta. *Pesquisa FAPESP, 25*(336), 54–57.

Balbachevsky, E. (2000). *A profissão acadêmica no Brasil: As múltipla facetas do nosso sistema de ensino superior*. FUNADESP.

Balbachevsky, E. (2008). Incentives and obstacles to academic entrepreneurship. In *University and development in Latin America* (pp. 23–42). Sense Publishing.

Balbachevsky, E. (2013). Academic research and advanced training: Building up research universities in Brazil. In *Latin's America's new knowledge economy: Higher education, government and international collaboration* (pp. 113–133). Sense Publishing.

Braun, D. (2003). Lasting tensions in research policy-making – A delegation problem. *Science and Public Policy, 30*(5), 309–321.

Castello, L., & Stewart, D. J. (2010). Assessing CITES non-detriment findings procedures for Arapaima in Brazil. *Journal of Applied Ichthyology, 26*(1), 49–56.

Castro, M. C. (2022). Improving health in the Amazon demands local involvement. *Nature Medicine, 28*, 435. https://doi.org/10.1038/s41591-022-01710-9

Ciarli, T., & Ràfols, I. (2019). The relation between research priorities and societal demands: The case of rice. *Research Policy, 48*(4), 949–967.

Digital Science. (n.d.). *Discover Dimensions the Most Advanced Scientific Research Database*. Digital Science. Retrieved October 24, 2023, from https://www.digital-science.com/product/dimensions/

Durham, E. R., & Gusso, D. A. (1991). Pós-graduação no Brasil: Problemas e perspectivas. Paper presented at the International Seminar on the tendencies of Brazilian post-graduation, MEC/CAPES: Brasilia, July 10–11, 28 pp.

FAPESP – Chamada para propostas FAPEAM FAPESP. (2020). https://fapesp.br/14005/chamada-de-propostas-fapeam-fapesp-2020

Gatti, L. V., Basso, L. S., Miller, J. B., et al. (2021). Amazonia as a carbon source linked to deforestation and climate change. *Nature, 595*, 388–393. https://doi.org/10.1038/s41586-021-03629-6

GEOCAPES. (n.d.). Retrieved October 24, 2023, from https://geocapes.capes.gov.br/geocapes/

Imbiriba, E. P. (2001). Production potential of pirarucu, Arapaima gigas, in captivity. *Acta Amazonica, 31*, 299–299.

Marcovitch, J., & Val, A. (2024). Bioeconomia para quem? bases para um desenvolvimento sustentável na Amazônia. Universidade de São Paulo. Escola de Comunicações e Artes. https://doi.org/10.11606/9786589321453

Merton, R. K. (1968). The Matthew effect in science: The reward and communication systems of science are considered. *Science, 159*(3810), 56–63.

Saes, M.S.M., Feitosa, E.R., Rocha, A.B., Bessa, J.D.R (2024) Vazio Institucional e Guardiões do Pirarucu do Amazonia. In Marcovitch, J. eds. Bioeconomia para quem? bases para um desenvolvimento sustentável na Amazônia. Universidade de São Paulo. Escola de Comunicações e Artes. https://doi.org/10.11606/9786589321453

Schwartzman, S. (1991). *A Space for Science: The development of the scientific community in Brazil.* Pennsylvania State University Press.

Thornton, P., Ocasio, W., & Lounsbury, M. (2012). *The institutional logics perspective: A new approach to culture, structure and Process.* Oxford University Press.

Van Eck, N. J., & Waltman, L. (2007). VOS: A new method for visualizing similarities between objects. In H.-J. Lenz & R. Decker (Eds.), *Advances in data analysis: Proceedings of the 30th annual conference of the German classification society* (pp. 299–306). Springer.

CHAPTER 10

Allies in the Combat Against Climate Change? The Role of Climate Change for University Alliances

Alina Felder-Stindt

Introduction

While it has primarily been internationalization processes that drive the formation of higher education institution (HEI) alliances for collaborations in teaching and research (Knight, 2014), the European Union (EU) has recently turned to HEI alliances as problem solving tools for grand societal challenges (Fuchs et al., 2023). The core initiative of the EU's agenda in this regard is the European Universities Initiative (EUI), which was originally launched as a pilot initiative in 2019 and by now is part of the EU's HE policy framework program Erasmus+ (Charret & Chankseliani, 2022; Cino Pagliarello, 2022; Gunn, 2020). With various HEI alliances having existed at border region and supranational levels before (Janczak, 2016; Trippl et al., 2015; Vukasovic & Stensaker, 2018),

A. Felder-Stindt (✉)
School of Economics and Political Science, University of St. Gallen, St. Gallen, Switzerland
e-mail: alina.felder-stindt@unisg.ch

J. Kekäle, R. Pinheiro (eds.), *Higher Education Policy for Tackling Climate Change*, https://doi.org/10.1007/978-3-031-78479-8_10

the landscape of HEI alliances in Europe is becoming more populated and diverse in terms of the alliances' objectives (Pinheiro et al., 2023).

Forming alliances among HEI has always been a political endeavour both at the level of (sub)national decision makers and of HEIs themselves. Alliance cooperation objectives result from a complex interplay of multiple levels, actors, and issues that depend on the political agenda of their source of funding (Knight, 2014; Pasqualoni & Scott, 2011). With climate change assuming a significant role in political agendas at (sub)national and supranational levels, it is only natural that issues of sustainability are impacting international cooperation in HE. The research question this chapter asks is: ***Why and to what extent do HEI alliances address climate change?***

This chapter contributes with a supranational policy perspective to the overarching research interest of this edited volume insofar the effects of political decision-making on HEIs' attitudes, policies, and initiatives towards CC mitigation. From the outlined stakeholders as per the editorial Chap. 1 of this volume, this chapter addresses primarily representatives from the first two groups, that is, state (politicians) and HEIs (academics). Both influence HE institution alliance formation with the explicit objective to mitigate CC. In addition, these two groups influence the formation and cooperation results of HEI alliances through an interplay of structural and relational mechanisms. These mechanisms relate to key external influences and issues to avoid (structural mechanism), on the one hand, and to key intentions and major assets in relation to CC mitigation (relational mechanism), on the other hand.

The study of HEI alliances against the backdrop of political agendas and different policy instruments to mitigate climate change is not only timely, but HE scholars have also neglected the role of climate change for HEI alliances so far has. Along with the empirical focus of this book, this chapter focuses on HEIs in the process of the sustainable and green transitions. More specifically, this contribution addresses how HEI alliances are affected by the impetus for a stronger contribution by HEIs to tackle climate change (Stein, 2024). The objective of this chapter is thus to investigate the role of climate change for the missions and cooperation results of HEI alliances in the European Education Area (EEA). For this purpose, this chapter develops a theoretical framework that draws on HE regionalization and Europeanization research. The framework serves the comparative analysis of HEI alliances that have received funding through the EUI.

After briefly reviewing existing scholarship in the next section, this chapter develops a theoretical framework that captures both internal and external perspectives on the role of climate change for the establishment and workings of HEI alliances. Having elaborated on the compared cases and the adopted methodology, the chapter then proceeds to the presentation of the empirical findings. Along with the broad conceptual framework adopted throughout this edited volume, this presentation regards drivers, steering mechanisms, outputs, and outcomes. While the first part of the analysis unravels why HEI alliances address issues of climate change mitigation and sustainability, in a second step the analysis unveils how climate change mitigation influences their concrete cooperation activities. Finally, and after having discussed the results of the analysis, the chapter concludes with an outlook as regards implications and future studies.

HEI Alliances and the Climate-HE Nexus

This chapter speaks to the scholarly communities that have analysed the formation, workings, and cooperation results of HEI alliances, on the one hand, and the role of climate change in HE and at HEIs, on the other hand. Bringing these research strands together is worthwhile, especially considering that there is little research to date concerning the thematic education and/or research foci of HEI alliances (Cojanu et al., 2016). While our knowledge of HEI alliance formation and its results is vast, existing research has largely failed to address the role of concrete thematic issues when HEI alliances are established and implemented.

Already prior to the EUI as the most recent EU policy instrument for HEI alliance formation, a wide range of HEI alliances populated the European HE landscape (Pinheiro et al., 2023). This is documented by research on the internationalization of HE (Stensaker & Fumasoli, 2017) and on interest representation in knowledge-related policy fields (Gornitzka, 2015; Gornitzka & Sverdrup, 2011; Vukasovic, 2017). What distinguishes HEI alliances from the ad hoc cooperation of scholars and staff is their strategic approach. The term "transnational strategic partnership" describes the collaboration between a minimum of two HEIs in two different countries, where cooperation goes beyond the department, programme, or individual faculty-member level and takes place at the organizational level (Beerkens, 2002; Karvounaraki et al., 2018). While in theory transnational strategic alliances ought to address all the different HEI missions and tasks, cooperation usually takes places in selected areas

(Fehrenbach, 2023). Regarding education, HEI alliances serve the development of joint programmes and curricula and student and staff mobility (Knight, 2014). Strategic partnerships also expand towards the research mission and help to flesh out joint research foci (Felder, 2020; Giband & Mary, 2018).

The scholarly literature on transnational strategic alliances has assessed alliance formation purposes both from an organizational viewpoint and from a public policy perspective. When analysed in their capacities to strategically advance internationalization, HEI alliances are discussed as tools to strengthen the profile of the respective member institutions (Chapman et al., 2014; Fumasoli et al., 2020). Analyses have also zoomed into the benefits of education and research collaborations for students and staff (Frame & Curyło, 2022). Researchers further conceptualize HEI alliances as interest groups and analyse their role in policy making processes (Vukasovic & Stensaker, 2018) and in European HE cooperation (Marques & Graf, 2023). Finally, there is a growing body of scholarship concerned with the role of HEIs in fulfilling political agendas such as those related to fostering European integration, economic competitiveness, and diplomacy (Felder & Tamtik, 2023; Gunn & Mintrom, 2022).

Research on the role of climate change for HEIs primarily concerns climate change education, that is, the extent to which climate change is being dealt with in the context of university programmes (Filho, 2010; Leal Filho et al., 2021). Universities have been found to develop and adopt curricula and pedagogical approaches "to educate students (and by extension society) about the imperatives of carbon neutrality and climate change mitigation and adaptation" (Leal Filho et al., 2021, p. 109). Moreover, research at the climate change-HE nexus has begun to turn to the adoption of carbon-neutral practices of HEIs such as travel policies for staff (Da Silva et al., 2023). The greening of educational content and institutional practices display long-term approaches, which stands in contrast to urgent political short-term mitigation targets.

What is lacking from the first relevant strand of research are systematic analyses on the role of concrete societal challenges and political agendas for the development and workings of HEI alliances. It is only recently that scholars address thematic profiles of HEI alliances. In so doing, Stensaker et al. (2023) focus on alliance formation and are less considerate of concrete cooperation results. The scope of analysis of the second relevant strand of research are individual HEIs even though HEIs increasingly engage in transnational cooperation. This gap reflects the literature on third missions where the focus has been on single institutions as well (Pinheiro et al., 2015). A notable exception is the study by Arnaldo Valdés and Gómez Comendador (2022) which analyses the contribution of EUI alliances to civic universities.

Higher Education Institution Alliances, Societal Challenges, and Europeanization

Along existing research, we may distinguish between the following purposes of HEI alliances: institutional positioning, collaboration in education and research, interest representation, and furthering policy objectives. Each purpose has a different relational angle and either concerns the relationship between different HEIs (i.e. positioning and collaboration) or the relationship between HEIs and policy makers (i.e. interest representation and policy objectives). For either of the relationships we find tensions between top-down pressures and bottom-up initiatives and between cooperation and competition. In bringing these two perspectives together, this chapter borrows from Europeanization research, which has conceptualized the relationship between top-down and bottom-up influences. Drawing on Europeanization approaches, this chapter proposes three mechanisms and three theory-driven expectations (see Table 10.1).

While changes in attitudes, public policies, and institutions through Europeanization are not unilateral, the observed complexity was usually reduced, which led to a focus on analysing the influence of European

Table 10.1 Theoretical Framework and Theory-Driven Expectations

	Cooperation objectives		*Cooperation results*
	Structural	*Relational*	*Cognitive*
Mechanism	Opportunity structures constrain/enable cooperation	Use of opportunity structures for positioning/influence	Adaptation of means/ends of cooperation
Expectation	HEI alliances focus on the issue of climate change due to funding instruments and/or societal expectations.	HEI alliances cooperate on the issue of climate change to strengthen their profile and/or exert influence.	Climate change shapes HEI alliances' cooperation activities and/or the nature and scope of cooperation.

Source: Author's own following Börzel and Risse (2003), Jacquot and Woll (2003), and Radaelli (2004)

policymaking on domestic policies and politics (Börzel & Risse, 2000). Europeanization literature proposing a top-down perspective is institutionalist, so that the adaptation pressure of Europeanization unfolds either vertically or horizontally, that is, through compliance or socialization (Bache, 2010; Radaelli, 2003). The two major new institutionalisms that find application to Europeanization research are rational-choice and sociological institutionalism. From the perspective of the former, Europeanization occurs as a reaction to EU opportunities or constraints. This reaction consists in "a cost-benefit analysis, where the cost of becoming Europeanized is outweighed by the benefits it brings, such as additional resources" (Huggins, 2018, p. 1266). Sociological institutionalism posits that Europeanization lets actors internalize EU rules, norms, and preferences, which are translated into the respective (sub)national context (Börzel & Risse, 2003; Mörth, 2003).

Since "domestic actors [...] may draw on Europe as a resource without specific pressure from Brussels" (Radaelli, 2004, p. 4), wider and notably sociological understandings of Europeanization have been developed (Büttner et al., 2022). Here, analyses do not only explain domestic change but also explain "how the norm developed on the EU level has been influenced by domestic debates, positions and power games between actors" (Saurugger, 2014, p. 184). Enriching institutionalist thinking with discourse and agency is at the heart of relational approaches towards Europeanization. From a relational perspective, EU-level opportunity structures are still a necessary condition for Europeanization to occur. Yet, Woll & Jacquot (2010, p. 116) postulate that change on the (sub)national level results from "social practices that seize the European Union as a set of opportunities". The seizing opportunities appear as strategic, cognitive, or legitimizing and rely on material (EU institutions, policy instruments, financing) and immaterial elements (discursive references, ideas, and the public sphere) (Jacquot & Woll, 2003, p. 6).

Both structural and relational approaches to Europeanization have developed ideas on its potential outcomes. Scholars adopting a structural approach have argued for inertia, accommodation, transformation, and/or retrenchment as possible results of Europeanization (Heritier & Knill, 2001; Risse et al., 2001). Relational Europeanization is not only expected to result in additional resources or reduced costs for (sub)national actors, but also in altered preferences and eventually in the constitution of new actors (Jacquot & Woll, 2003, p. 5). While there is a variety of labels to capture change through Europeanization, there is consensus that the effects of Europeanization are either superficial or go much deeper in replacing (sub)national structures, practices, and rationales (Risse et al., 2001; Saurugger, 2014). The idea of different depths of change resonates with the policy change and learning literatures and has been coined as cognitive Europeanization (Liebert & Calloni, 2003). The framework adopted in this chapter combines structural, relational, and cognitive approaches to Europeanization as to analyse how climate change influences HEI alliances' objectives and cooperation results (Table 10.1). This combination allows us to formulate theoretical expectations for top-down and bottom-up influences on cooperation objectives and for the depth of cooperation results. From a structural perspective, we expect that HEI alliances address the issue of climate change in their cooperation because of funding instruments and/or societal expectations. A relational logic manifests itself in HEI alliances focusing on climate change to strengthen the profile of the involved HEIs and/or in HEI alliances aiming to exert political influence. Along the cognitive mechanism, we expect that the focus on climate change is reflected in concrete cooperation activities and/or even influences the way in which HEIs cooperate with one another.

The three mechanisms and related theoretical expectations speak to the three analytical dimensions set up by the editors at the onset (see Chap. 1) of this volume—drivers, steering mechanisms, and effects. The question of *why* HEI alliances address the issue of climate change relates to identifying endogenous and exogenous drivers, which in the scope of our analysis pertains to manifestations of relational and structural Europeanization. To analyse the extent to which HEI alliances steer cooperation on issues of climate change it is useful to distinguish between cooperation outputs and outcomes. This aspect is closely related to assessing cognitive Europeanization, which either results in new activities, such as joint educational offers, or leads to novel understandings on how to implement such activities.

Cases and Methodology

The analysis in this chapter focuses on HEI alliances that have received funding through the European Universities Initiative (EUI). The latter began as a pilot led by the European Commission's Directorate General for Education and Culture (DG EAC) in late 2018 and by now builds one strand of the Erasmus+ program with optional top-up funding through the EU's research funding programme, Horizon Europe. After a fifth round of calls for alliance applications (2019, 2020, 2022, 2023 and 2024), a total of sixty-four alliances have received funding. Starting in the third round of calls, in 2022, not only new but also already established alliances could apply for funding.

The EUI serves to establish "transnational alliances of higher education institutions developing long-term institutional structural and strategic cooperation [...] and aiming to achieve sustainability of their cooperation" (European Commission, 2021, p. 5). Through transnational cooperation, European Universities are supposed to promote European values and identity and to increase the competitiveness of European HE. To achieve this aim European Universities (1) develop "an integrated long-term joint strategy for education with, where possible, links to research and innovation, and service to society", (2) establish a European HE inter-university campus, and (3) follow a challenge-based interdisciplinary approach to learning (European Commission, 2019, pp. 125–126).

The collected and analyzed data for this chapter includes the first to fourth call for EUI alliances and thus concerns fifty out of the overall sixty-four EUI alliances. According to their factsheets, out of the 50 EUI alliances that were established by 2023 twenty-seven address issues of climate change and sustainability. While documents such as mission statements and information on alliance webpages were considered for these twenty-seven case alliances, interviews were conducted with one representative each from twelve alliances. Out of the twelve interviewed alliances, five received funding in the first and third rounds and seven are beneficiaries of the second and fourth rounds. 4EU+, CHARM-EU, ECIU University, EPICUR, and Una Europa receiving funding through the calls in 2019 and 2022; CIRCLE U., EU3DRES2, ENGAGE.EU, ENHANCE, ENLIGHT, Eut+, and Transform4Europe receiving funding through the calls in 2020 and 2023. The interview respondents were selected along their responsibility for the themes of climate change and sustainability in their alliance. Depending on the alliance governance structure, these responsibilities lie with employees in the alliance central offices or with members of the academic or administrative staff at one of the alliance's member institutions. The qualitative semi-structured interviews included questions about alliance formation and objectives and the results of cooperation with a particular focus on the role of climate change and sustainability for the alliances' objectives and cooperation (actual) results. The interviews were conducted between August and November 2023 in either English or German and lasted for an hour each.

The collected documents and verbally transcribed interviews underwent a thematic analysis (Braun & Clarke, 2022). The thematic coding framework to analyse the data is based on the reviewed literature and the theoretical framework and on data familiarization. Applying the coding framework to the gathered data helped to identify and review potential themes in a first step and to define and specify themes in a second step. The results of this process are documented in the following section and for the purpose of illustrating the themes draw on quotations from documents and the conducted interviews.

Analysis: The Role of Climate Change for Higher Education Institution Alliances

As alluded above, the first part of the analysis concerns the question of *why* HEI alliances address the issue of climate change. This part of the analysis relates to the interplay of structural and relational mechanisms and thus to identifying endogenous and exogenous drivers. In a second step, the analysis is concerned with the question of *how* HEI alliance cooperation deals with climate change. This part of the analysis relates to cognitive mechanisms and distinguishes between cooperation outputs and outcomes.

Why HEI Alliances Address Climate Change

The thematic foci of the analysed HEI alliances are the result of mapping and brainstorming exercises where the profiles and activities of alliance members are surveyed and working groups composed of diverse types of HEI staff in research, teaching, and administration discuss shared topics. Based on the mapping and/or the staff drafts for focus area activities, the central governing body of an alliance—usually a council of HEI presidents—decides on thematic foci of alliance cooperation. Considering both exogenous and endogenous drivers, this section shows how climate change results as a theme for HEI alliance cooperation (see Table 10.2).

Regarding *exogenous drivers*, the analysis shows that funding stipulations ask for a contribution of HEIs to solve societal challenges. While the first two calls (2019 and 2020) remained open about the societal challenge an alliance will work on to create "innovative solutions adaptable to different regions in Europe" (European Commission, 2019, p. 126), the later calls in 2022 and 2023 spell out the areas of societal challenges. Since

Table 10.2 Endogenous and Exogenous Drivers Behind HEI Alliance Objectives

Endogenous drivers	*Actors*	*Exogenous drivers*	*Actors*
Complementarities of existing education/research foci at alliance member HEIs	Individual scholars/ teaching staff	Funding requirements	European Commission Governments
Pre-existing cooperation foci among HEIs joining an alliance	Network champion(s)	Political and societal demands	Student body and/or society

Source: Author's own

then, the integrated long-term strategy for European Universities is supposed to be "responsive to the digital and green transition and key socio-economic challenges, while remaining committed to excellence" (European Commission, 2021, pp. 5–6). Moreover, since 2022 the second objective to establish inter-university campuses refers to green and digital skills (European Commission, 2021, p. 6) and the passage detailing the expected impact of European Universities mentions HEI alliances as "actors of change for the digital and green transitions" (European Commission, 2021, p. 9). The background of spelling out the contribution of HEIs to sustainability are several EU conclusions and communications that stress the role of skills for "sustainable competitiveness, social fairness and resilience" (European Commission, 2020, p. 1) and for "the technological and green transition in support of inclusive and sustainable growth" (Council of the European Union, 2019, p. 1). The first sentence of the European Strategy for Universities refers to climate change as one of the major challenges facing Europe. The strategy emphasizes how the "higher education sector has an essential role to play in Europe's post-pandemic recovery and in shaping sustainable and resilient societies and economies" (European Commission, 2022, p. 2).

Both analysed interviews and documents show that the notion of HEIs as societal problem solvers influences the objectives of an HEI alliance. One interviewee explained this logic as follows:

> *Such applications always reflect the requirements. Because if you do something different, then it does not get funded and then you've made the effort for nothing. It's the call and response, you tell me what you want, I give you what you want and then I get what I want. That's always the case and you have some leeway for creative implementation.* (Interviewee Circle U.)

The analysed mission statements of HEI alliances closely reflect the demands of the EU funding opportunity they benefit from. This becomes evident in the recurrent use of the following terms from the EU's call in the alliance mission statements: "challenge-based learning"; "mission-oriented research"; "human-centred and cross-sectoral innovation"; "engaged knowledge exchange"; "responsible changemakers"; etc. While it is no surprise that alliances' mission statements strongly reflect EU speech, also because the mission statements are required deliverables, the interviews show that political demands travel further than framing exercises in documents targeting the European Commission. Certain interviewees adhered to notions of the "*university of the future*" (Interviewee, ECIU) with a "*future-proof education and R&I agenda*" (Interviewee ENLIGHT). One interviewee explained that the joint teaching offer of the alliance should enable "*students to become citizens who are able to face and to contribute to solving the challenges we are facing in Europe and beyond*" (Interviewee 4EU+).

Insofar the demands of policy makers, interviewees at EPICUR and EU3DRES2 explained that the student body also voices thematic preferences which are linked to the growing public awareness about sustainability. Societal demands differ from funding requirements. Since students and HEI staff are constitutive parts of a society that is increasingly attentive to climate protection, their demands are intricately linked to norms which influence HEI alliance cooperation. A statement from an ENHANCE representative vividly reflects this aspect:

> *It is not just about developing teaching offers for students, but it is also about giving them the feeling that what we are doing as an alliance has an overriding meaning. This means conveying that they are at a university that wants to shape the future. This part of the work of an alliance brings the question; how do we see ourselves as universities within society?*

The establishment of an alliance requires HEIs to reconsider their role in society. At the same time, the interviewee at ENHANCE outlined that with regards to solutions for climate change, society also needs to be ready to use them and that "*in fact, when it is about sustainability, we are further from where we would want to be as a society*".

Concerning *endogenous drivers*, the interviews reflect that beyond the attention towards political and societal expectations HEI alliances themselves recognize that climate change is a transborder issue which invites collaboration and the need for a joint approach. In the analysed cases, the theme of climate change resulted as an area for cooperation when there has been complementary education and/or research foci at an alliance member institution in this regard. One interviewee from the Una Europa case explained that sustainability has resulted as one of the alliances' focus areas because, it was already addressed by the nine member HEIs and these would always try to bring their thematic foci into their international cooperation outlets. When asked about the difference the EUI makes to tackle themes related to climate change, another interviewee referred to the alliance as "*a way to disseminate the topic of sustainability to a broader number of students within different countries*" (Interviewee ENLIGHT).

Such responses reveal a *relational logic* that lets HEIs make use of the EUI to increase their capacities. Against this backdrop, HEI alliances provide a framework to pool resources such as their teaching offerings and/or to jointly acquire new funding. Various interviewees referred to cases where one member HEI had resources and structures at its disposal that built the basis of alliance cooperation. Examples include existing study programmes with a sustainability focus, the infrastructure to set up MOOCs, and sustainable travel policies (Interviews ECIU University, Eut+, Transform4Europe). Consequently, and in those cases where climate change features in existing activities of the majority of HEIs in an alliance, the partners would want to benefit from a thematic overlap and/or profit from more advanced partners. An interviewee at EU-CONEXUS indicated that in the area of climate education the alliances' "*universities come in with the different disciplinary profiles, so that the students get a whole spectrum that they cannot get from one partner university only*".

Increased capacities to work on a certain theme may also feed into the pro-active approach of certain alliances where a thematic cooperation focus serves to influence policymaking (agenda setting). This is because the answers of an HEI alliance to societal challenges have already been lifted to the European level and will not "*only be the expression of a single university*" (Interview 4EU+). Expressions of the ambition to exert influence on institutional practices beyond one alliance and on policymaking take the form of whitepapers, open letters, research papers, etc. Next to different policy papers, the Circle U. alliance for example has published a conceptual paper on sustainable education. Such papers are coordinated and published by the Circle U. Think and Do Tank which "*engages with students and external stakeholders with the aim to influence the policy agenda on higher education*" (Interviewee Circle U.).

Overall, the data suggest that it is the combination of internal and external influences which leads to a thematic focus on climate change–related issues. The interrelationship between endogenous and exogenous drivers becomes evident in the following statement from an interviewee at ENLIGHT where climate change is one "cross-cutting challenge domain":

> *We thought about what the alliance's strengths in this area are. But also, of course, considered what will be expected from the experts and they will of course expect something on the subject of protection, climate change and climate adaptation.*

According to one interviewee at Una Europa, the topics of climate change and sustainability were chosen for alliance cooperation because of "*their societal relevance, but also because the exchange of best practices at the inter-institutional level is worthwhile*". Since the activities of the assessed HEI alliances would not have been established without the EUI and EU funding, certain constraints for cooperation are inevitable. One interviewee at Circle U. elaborated on this:

> *We have committed ourselves to certain things with the application and you have to swear in the people who will later be involved in the project and didn't help develop it. Nobody has anything against tackling climate change, that's not the problem, but the question of implementation and what exactly should happen there or the scope for action is extremely limited by the funding application.*

Overall, the analysis points to structural and relational mechanisms jointly influencing the thematic orientation of HEI alliances. A thematic focus on climate change results from respective funding instrument objectives and societal demands (structural mechanism) and from complementary education and research foci at alliance HEIs (relational mechanism).

How HEI Alliances Address Climate Change

There are four areas in which the analysed HEI alliances cooperate in issues related to climate change. These are teaching, research, innovation, and institutional practices. In each of these areas we may distinguish between shorter-term outputs and longer-term outcomes (see Table 10.3). In the majority of the analysed alliances, the issue of climate change

Table 10.3 Cooperation Results of HEI Alliances Focusing on Climate Change

	Teaching		*Research*	*Innovation*	*Institutional practice*	
Outputs	Workshops Summer schools Micro-credentials MOOCs	PhD summer schools Thematic conferences	Joint research projects Visiting professorships	Social innovation projects	Events with practitioners Staff training for innovative teaching	Alliance mission statements
Outcomes	Joint study programmes	Access to courses across HEIs	Joint research areas	Knowledge hubs	Working groups	Sustainable travel policies

Source: Author's own

features as a dedicated focus area for cooperation and/or is part of a focus area with a broader sustainability framing. The purpose of these focus areas is to thematically streamline joint teaching and research activities and the interaction of HEIs with their societal contexts. Beyond teaching and research, climate change is also addressed with regards to institutional practices.

The cooperation in teaching leads to joint education formats, where specific or all partner institutions of an alliance at least establish a short-term course offer which is accessible for students across the alliance. This offer takes the shape of workshops and summer schools, and what is particular to the European Universities is the focus on module-oriented teaching offers such as specialization modules, minor programmes, and micro credentials, a path which according to the interviews certain HEIs have only embarked upon due to their alliance (Interview EU-CONEXUS). The outcomes in teaching that have the potential to last beyond the funding of the HEI alliances through the EUI include the opening of individual HEIs' offers to students from other alliance members and the development of joint study programmes. In contrast to the large body of students and staff at one alliance, the joint teaching offers are usually only accessible to a limited number of participants. This is particularly the case for workshops and summer schools and less the case for joint study programmes.

Regardless of their length, the joint teaching formats of the analysed HEI alliances emphasize mobility and challenge-based learning. The mobility criterion is due to the Erasmus+ program background of the EUI, which ever since it was created aims to increase the mobility of students and teaching staff across Europe. The purpose of joint courses on issues related to climate change is to "*develop concrete solutions to address specific sustainability challenges*" (Interviewee CHARMEU). To fulfil this objective, the final term of the Master's programme *Global challenges for Sustainability* of the CHARMEU alliance, for example, involves a thesis project where students do not only work together with their academic supervisors, but also with practice partners such as civil society organizations. The result of the thesis ought to be a solution to a challenge related to sustainability.

The challenge-based understanding of teaching is closely related to the research activities of an alliance, which are less encouraged through the EUI funding scheme, but still take place in the majority of the alliances given the strategic importance of research in HEI alliances. Here the identified shorter-term activities include joint research projects and visiting professorships, whereas PhD summer schools are at the crossroads of cooperation in teaching and research. As outlined above, the thematic guidance of HEI alliance cooperation is usually steered through so-called flagship areas, such as *Climate Change* (ENLIGHT), *Environmental Transitions* (4EU+), and *Sustainable Society* (EuroTech). Research on cross-border cooperating HEIs shows that if thematic focus areas are maintained after the project nature of the analysed alliances, they may feed into the establishment of joint research areas. One format of cooperation at the crossroads of teaching, research, and outreach are ***knowledge hubs***. The Circle U. alliance has established such hub "to explore how universities can contribute with knowledge and solutions to the pressing issues of climate change and examine the implications of the European Green Deal" (Mission Statement Circle U.).

The objective of increasing inter-institutional mobility conflicts with objectives related to more sustainable institutional practices. The development of sustainable institutional practices is the task of administrative staff, which are connected with each other through working groups. Examples include the *Green Office Network* of the ENLIGHT alliance which "connects respective green offices of all partners to work on sustainable solutions and tools to reduce the carbon footprint of internationalization" (Mission Statement ENLIGHT). In their quest to make campuses, studying and working in academia greener, around half of the analysed HEI alliances are working on sustainable travel policies and sustainability strategies. Examples include the *Task Force for Sustainability and Climate Protection* in the Una Europa alliance, which develops a sustainability strategy. Likewise, EU CONEXUS has mapped institutional practices with the objective to make the alliance members' campuses greener. From here, a joint action plan was developed and "*depending on the institutional advancement, universities can pick certain actions that are proposed in the action plan*" (Interviewee EU-CONEXUS).

The identified cooperation results result from a combination of ex structural and relational mechanisms. To further HEI alliance cooperation in teaching, EU funding guidelines are crucial. Notably, the primary purpose of the alliances funded through the EUI is the education of students across borders through innovative teaching formats. The fulfilment of this objective is measured through quantifiable indicators such as the number of conducted events and the respective participants. To complement the instructions as given through funding requirements, many alliances have opted to award seed funding to incentivize bottom-up activities such as thematic exchanges among students and staff and for the purpose of developing joint projects.

While short-term cooperation formats have existed in all the analysed cases, the establishment of joint study programmes and joint research areas was found to be less common. The interviews reveal that there are varied reasons why deeper changes and long-term results of alliance cooperation in the field of climate change are less salient. Next to the different institutional profiles and sizes among partners, several interviewees pointed to the differences in national legislation, which raises obstacles for cross-border cooperation. While these are challenges which HEI alliances traditionally face, an HEI alliance is more likely to focus on climate and sustainability themes if these are a national priority and if partner HEIs are already working on climate change and sustainability topics. At the same time, different political and institutional attention towards these topics can pose challenges for cooperation.

Even if climate change mitigation is a priority at political and institutional levels, the translation of this priority requires resources and institutional capacity. When interviewees state that "*there is this impression that some people in Brussels are counting on the enthusiasm and the conviction at the institutional level and of course at the alliance level*" (Interviewee 4EU+), this reveals that cooperation requires more resources than those which the EUI funding mechanism provides and finances. The provided funding is perceived as too little in comparison to what is asked from the alliances. The interviewees also pointed to the differences in remuneration, with only certain HEIs acknowledging the extra hours employees spent working for the alliance. In their quest for more resources almost all alliances have expanded their consortia when applying for a second round of funding as this meant additional financial resources for the alliance. Getting existing alliances to re-apply with an extended membership was a clear objective of the European Commission. At the same time, more HEIs in an alliance increase national and institutional diversity, which in turn brings novel cooperation challenges.

Finally, the focus of HE alliance cooperation around climate change depends on time. One interviewee at 4EU+ explained "*we are for now at a stage where each institution tries to find his place also on this topic*". Being too recent a topic, cooperation results that concern climate change tightly connect to the length of alliance cooperation. Several interviewees emphasized that their alliance is young and that they have only just started cooperating, which means that the major result is the establishment of a structure and framework for cooperation, from which concrete thematic cooperation will follow. The cooperation of the European University alliances so far only spans over short periods of time. It remains an open question how much the cooperation in the alliances will change understandings of HEIs and their roles insofar teaching, research, and their impact in society, including as regards critical issues like climate change.

Conclusive Thoughts: Allies Yes, But!

The purpose of this chapter was to examine the role of climate change for the mission and cooperation results of HEI alliances. For that purpose, a theoretical framework drawing on Europeanization research was developed and operationalized; combining structural, relational, and cognitive mechanisms, along which intrinsic motivations and external pressures were thought to influence the thematic focus of HEI alliance cooperation. The analysis found that issues of climate change and sustainability feature in HEI alliance missions due to the profiles of the involved HEIs but also because of societal demands and funding stipulations. This confirms Europeanization research which has established that structural and relational mechanisms go hand in hand (Börzel & Risse, 2003; Radaelli, 2004). Regarding cooperation results, we found that HEI alliances are primarily successful in incorporating climate change mitigation in delivering short-term joint teaching formats and in bringing researchers with shared agendas and interests together. What is more, HEI alliances serve to exchange best practices to make campuses and university life more sustainable. The achievement of longer-term outcomes such as joint study programmes proves and sustainability policies such as concerning student and staff mobility to be more complex and difficult because of diverging HE systems and regulations. This is in line with cognitive Europeanization research which has found that deeper changes in practices and policies only unfold over longer periods of time (Liebert & Calloni, 2003).

The findings in this chapter, on the one hand, confirm that European HEIs are increasingly responsive to societal and political developments and thus extend their tasks beyond education and research (Fuchs et al., 2023; Pinheiro et al., 2015). HEIs pursue alliances and collaboration based on staffs' expertise and preferences (Fehrenbach, 2023). As a result, HEIs usually are part of multiple alliances which fulfil different purposes (Stensaker & Fumasoli, 2017). While the results of the study at hand suggest that it is very much HEIs staff themselves who see the necessity to incorporate issues of climate change and sustainability in their daily work, incentives such as prestige and funding cannot be neglected.

On the other hand, the findings imply that next to unifying HEI and their members in the fight against climate change, they also serve the strategic positioning of partner institutions. Choosing climate change mitigation as an alliance mission serves strategic purposes such as to secure funding or to strengthen one's competitive or market positioning nationally and globally (Fumasoli et al., 2020). Practitioners tasked with establishing HEI alliances and implementing their activities consider political agendas as to receive financial and political support for their cooperation (Vukasovic & Stensaker, 2018). To what extent this will lead to window-dressing or symbolic behaviour of HEI alliances remains an open question (Stensaker et al., 2023). What is certain is that the agreement on the need to tackle climate change and to engage in collaborations in teaching and research must not necessarily translate into a shared effort for the entire HE sector (Stein, 2024). Instead, opportunistic behaviour by HEIs in acquiring support for their cooperation is likely, especially considering that the funding horizon of the EUI is short-term (Karvounaraki et al., 2018) and that the European Commission incentivizes opportunistic behaviour through additional resources in case of alliance expansion.

Consequently, when theorizing about the formation and workings of cross-border HEI cooperation, one needs to incorporate specific policy issues more strongly and examine their enabling and/or constraining role (Chapman et al., 2014; Felder, 2020; Knight, 2014). Since HEI alliance formation usually builds on pre-existing ties among HEIs (Charret & Chankseliani, 2022) and in some cases the newly established alliances even rely on pre-existing alliances (e.g. ECIU University, EPICUR), the findings for the analysed cases may be tested for other cases of HEI alliances. Future analysis could adopt the theoretical framework of this chapter to analyse how the tackling of societal challenges such as climate change is reflected in the transnational alliances existing outside the European Universities scheme and border-region HEI alliances. Moreover, developed theoretical and empirical arguments may as well be tested for further societal issues that demand the contribution of HE institutions and their alliances, such as the digital transformation.

References

Arnaldo Valdés, R. M., & Gómez Comendador, V. F. (2022). European Universities Initiative: How Universities May Contribute to a More Sustainable Society. *Sustainability, 14*(1), 471.

Bache, I. (2010). *Europeanization and multilevel governance: Cohesion policy in the European Union and Britain. Governance in Europe.* Rowman & Littlefield.

Beerkens, E. (2002). International inter-organisational arrangements in higher education: Towards a typology. *Tertiary Education and Management, 8*(4), 297–314.

Börzel, T. A., & Risse, T. (2000). When Europe hits home: Europeanization and domestic change. *European Integration Online Papers, 4*(15).

Börzel, T. A., & Risse, T. (2003). Conceptualizing the domestic impact of Europe. In K. Featherstone & C. M. Radaelli (Eds.), *The politics of Europeanization* (pp. 57–80). Oxford University Press.

Braun, V., & Clarke, V. (2022). *Thematic analysis: A practical guide.* Sage.

Büttner, S. M., Eigmüller, M., & Worschech, S. (Eds.). (2022). *Sozialwissenschaftliche Einführungen: Vol. 3.* (Sociology of Europeanization). De Gruyter Oldenbourg.

Chapman, D. W., Pekol, A., & Wilson, E. (2014). Cross-border university networks as a development strategy: Lessons from three university networks focused on emerging pandemic threats. *International Review of Education, 60*(5), 619–637.

Charret, A., & Chankseliani, M. (2022). The process of building European university alliances: A rhizomatic analysis of the European Universities Initiative. *Higher Education, 86*(1), 21–44.

Cino Pagliarello, M. (2022). Higher education in the single market between (trans)national integration and supranationalisation: Exploring the European Universities Initiative. *Journal of European Integration, 44*(1), 149–164.

Cojanu, V., Gavris, A., & Robu, R. G. (2016). In search of emerging polities: Thematic agendas of selected European cross-border cooperation structures. *Romanian Journal of European Affairs, 16*(1), 57–71.

Council of the European Union. (2019, November 18). Council conclusions on the key role of lifelong learning policies in empowering societies to address the technological and green transition in support of inclusive and sustainable growth: (2019/C 389/07).

Da Silva, L. A., de Aguiar Dutra, A. R., & de Andrade Guerra, J. B. S. O. (2023). Decarbonization in higher education institutions as a way to achieve a green campus: A literature review. *Sustainability, 15*(5), 4043.

European Commission. (2019). Erasmus+ Programme Guide. https://erasmus-plus.ec.europa.eu/sites/default/files/erasmus-plus-programme-guide-2019_en_1.pdf

European Commission. (2020, July 1). Communication from the Commission to the European Parliament, the Council, the European Economic and Social Committee and the Committee of the Regions European Skills Agenda for sustainable competitiveness, social fairness and resilience: COM(2020) 274 final.

European Commission. (2021). Erasmus+ Programme (ERASMUS) Call for proposals Partnerships for Excellence—European Universities -ERASMUS-EDU-2022-EUR-UNIV [EU Grants: Call document (ERASMUS): V1.0 - 30.11.2021].

European Commission. (2022, January 18). Communication from the Commission to the European Parliament, the Council, the European Economic and Social Committee and the Committee of the Regions on a European strategy for universities: COM(2022) 16 final.

Fehrenbach, H. (2023). *Transnational alliances in higher education.* Springer Fachmedien Wiesbaden.

Felder, A. (2020). Crossing borders in higher education? A comparative case study at the intersections of EU regional and higher education policies. *Eastern Journal of European Studies, 11*, 28–45.

Felder, A., & Tamtik, M. (2023). Federal servants of inclusion? The governance of student mobility in Canada and the EU. *Politics and Governance, 11*(3), 251–263.

Filho, W. L. (2010). Climate change at Universities: Results of a world survey. In W. Leal Filho (Ed.), *Climate change management. Universities and climate change* (pp. 1–19). Springer.

Frame, A., & Curyło, B. (2022). Bringing Erasmus home: The European universities initiative as an example of 'Everyday Europeanhood'. *Journal of Contemporary European Studies, 32*(2), 334–349.

Fuchs, L., Cuevas-Garcia, C., & Bombaerts, G. (2023). The societal role of universities and their alliances: The case of the EuroTeQ Engineering University. *Tertiary Education and Management, 29*(3), 263–277.

Fumasoli, T., Barbato, G., & Turri, M. (2020). The determinants of university strategic positioning: A reappraisal of the organisation. *Higher Education, 80*(2), 305–334.

Giband, D., & Mary, K. (2018). Territorial cross-border cooperation in higher education. A case study of the eastern Pyrenean border. *Documents D'anàlisi Geogràfica, 64*(3), 587–601.

Gornitzka, Å. (2015). 'all in'? Patterns of participation in EU education policy. In M. Souto-Otero (Ed.), *Evaluating European education policy-making: Privatization, networks and the European Commission* (pp. 92–123). Palgrave Macmillan.

Gornitzka, Å., & Sverdrup, U. (2011). Access of experts: Information and EU decision-making. *West European Politics, 34*(1), 48–70.

Gunn, A. (2020). The European universities initiative: A study of alliance formation in higher education. In A. Curaj, L. Deca, & R. Pricopie (Eds.), *European higher education area: Challenges for a new decade* (pp. 13–30). Springer International Publishing.

Gunn, A., & Mintrom, M. (2022). *Public policy and universities: The interplay of knowledge and power. Cambridge elements. Elements in public policy.* .

Heritier, A., & Knill, C. (2001). *Differential responses to European policies: A comparison.* Max-Planck Project Group Preprint 2000/7.

Huggins, C. (2018). Subnational government and transnational networking: The Rationalist logic of local level Europeanization. *Journal of Common Market Studies, 56*(6), 1263–1282.

Jacquot, S., & Woll, C. (2003). Usage of European integration—Europeanisation from a sociological perspective. *European Integration Online Papers, 7*(12), 1–18.

Janczak, J. (2016). Cross-border universities as a field of Europeanization of higher education in the EU. In B. Przybylska-Maszner, M. Musia-Karg, & T. Brańka (Eds.), *Double reunification through the European Union's Education Policy* (pp. 37–50). Faculty of Political Science and Journalism Press.

Karvounaraki, A., Subramaniam, S., Hristov, H., Ojala, T., Jonkers, K., Huisman, J., & Goenaga, B. X. (2018). *Mapping of European transnational collaborative partnerships in higher education.* JRC Science for Policy Report.

Knight, J. (2014). Three generations of crossborder higher education: New developments, issues and challenges. In B. Streitwieser (Ed.), *Oxford studies in comparative education: Vol. 23, 2. Internationalisation of higher education and global mobility* (pp. 43–58). Symposium Books.

Leal Filho, W., Sima, M., Sharifi, A., Luetz, J. M., Salvia, A. L., Mifsud, M., Olooto, F. M., Djekic, I., Anholon, R., Rampasso, I., Kwabena Donkor, F., Dinis, M. A. P., Klavins, M., Finnveden, G., Chari, M. M., Molthan-Hill, P., Mifsud, A., Sen, S. K., & Lokupitiya, E. (2021). Handling climate change education at universities: An overview. *Environmental Sciences Europe, 33*(1), 109.

Liebert, U., & Calloni, M. (Eds.). (2003). *Gendering Europeanisation* (Series Multiple Europes: Vol. 19). PIE Lang.

Marques, M., & Graf, L. (2023). Pushing boundaries: The European universities initiative as a case of transnational institution building. *Minerva, 62*(1), 93–112.

Mörth, U. (2003). Europeanization as interpretation, translation, and editing of public policies. In K. Featherstone & C. M. Radaelli (Eds.), *The politics of Europeanization* (pp. 159–178). Oxford University Press.

Pasqualoni, P. P., & Scott, A. (2011). Positioning higher education in the imagined histories, geographies and futures of a European cross-border region. *Built Environment, 37*(3), 336–351.

Pinheiro, R., Gänzle, S., Klenk, T., & Trondal, J. (2023). Unpacking strategic alliances in European higher education. *Tertiary Education and Management, 29*(3), 213–228.

Pinheiro, R., Langa, P. V., & Pausits, A. (2015). The institutionalization of universities' third mission: introduction to the special issue. *European Journal of Higher Education, 5*(3), 227–232.

Radaelli, C. M. (2003). The Europeanization of public policy. In K. Featherstone & C. M. Radaelli (Eds.), *The politics of Europeanization* (Repr, pp. 27–56). Oxford University Press.

Radaelli, C. M. (2004). Europeanisation: Solution or problem? *European Integration Online Papers, 8*(16).

Risse, T., Cowles, M. G., & Caporaso, J. (2001). Europeanization and domestic change: Introduction. In M. G. Cowles, J. Caporaso, & T. Risse (Eds.), *Transforming Europe* (pp. 1–20). Cornell University Press.

Saurugger, S. (2014). Europeanisation in Times of Crisis. *Political Studies Review, 12*(2), 181–192.

Stein, S. (2024). Universities confronting climate change: Beyond sustainable development and solutionism. *Higher Education, 87*(1), 165–183.

Stensaker, B., & Fumasoli, T. (2017). Multi-level strategies in universities: Coordination, contestation or creolisation? *Higher Education Quarterly, 71*(3), 263–273.

Stensaker, B., Maassen, P., & Rosso, A. (2023). The European University Initiative—Investigating alliance formation and initial profile developments. *Tertiary Education and Management, 29*(3), 229–243.

Trippl, M., Sinozic, T., & Lawton Smith, H. (2015). The role of universities in regional development: Conceptual models and policy institutions in the UK. *Sweden and Austria. European Planning Studies, 23*(9), 1722–1740.

Vukasovic, M. (2017). Stakeholder organizations in the European higher education area: exploring transnational policy dynamic. *Policy and Society, 36*(1), 109–126.

Vukasovic, M., & Stensaker, B. (2018). University alliances in the Europe of knowledge: Positions, agendas and practices in policy processes. *European Educational Research Journal, 17*(3), 349–364.

Woll, C., & Jacquot. S. (2010). Using Europe: Strategic action in multi-level politics. *Comparative European Politics, 8*(1), 110–126.

CHAPTER 11

Universities and Climate Change: One Step Back and Two Steps Forward?

Jouni Kekäle and Rómulo Pinheiro

Introduction

The primary aim of this edited volume was to investigate the role of universities and other types of higher education institutions (HEIs) in tackling climate change (CC), by paying considerable attention to the role played by government policy and other types of formal and informal mechanisms pressuring HEIs to respond accordingly. Our intention was to consider drivers, mechanisms, and effects resulting from the complex interplay between HE—including the domains of science, research, and innovation—and policy initiatives in particular. In the introductory chapter we identified and discussed the role played by key stakeholders in the

J. Kekäle (✉)
School of Educational Sciences and Psychology, University of Eastern Finland, Joensuu, Finland
e-mail: jouni.kekale@uef.fi

R. Pinheiro
Department of Political Science and Management, University of Agder, Kristiansand, Norway
e-mail: romulo.m.pinheiro@uia.no

J. Kekäle, R. Pinheiro (eds.), *Higher Education Policy for Tackling Climate Change*, https://doi.org/10.1007/978-3-031-78479-8_11

process, what we term as "the key enablers" in green transitions (Table 1.1). The volume's rich empirical findings lend support to the notion that HEIs' roles in CC mitigation are part and parcel of a larger (eco)system of influences involving a multiplicity of actors, logics, values, and strategic aspirations.

First, HEIs are for the most part publicly funded and politically steered. Therefore, HEIs' contributions to CC inevitably occur in the context of the broader policy framework, with national and supranational (such as EU) policies affecting their work and roles in CC mitigation. Second, as pointed out in the opening chapter of this volume, companies, investors, policy makers, and citizens all affect the CC policy process (from agenda setting to implementation to its evaluation stages) alongside its mitigation. HEIs' contributions (or at least the outcomes of their CC-related activities) to the overall mitigation of CC depends heavily on these parties. Third, the public at large may either support or neglect any solution for CC mitigation. Public demand is a strong force which steers markets (companies and investors) and politicians alike. Thus, we argued at the onset that HEIs' roles in CC mitigation cannot be understood without considering this broader context, as visualised in a simplified manner in the introduction (Fig. 1.2).

Impact and mitigation of CC are requested from HEIs (Back to Earth, 2024), but, as suggested in Chap. 1, impact is not a one-way street (linear process) from HEIs to society, nor does it depend solely on HEIs, but other players also affect the outcomes, both tangible (quantifiable) and intangible ones. This can be seen as a complex system of interactions (see Chap. 1). In other words, the overall findings of the volume are aligned with earlier inquiries by the United Nations CC panel suggesting that:

> Successful adaptation [to CC but also other wicked issues facing human kind] not only depends on governments but also on the active and sustained engagement of stakeholders, including local communities, national, regional, multilateral and international organizations, public and private sectors, civil society and other relevant actors, as well as an effective management of knowledge. (UNFCCC)[1]

[1] Online at: https://unfccc.int/topics/adaptation-and-resilience/the-big-picture/introduction

HEIs can provide knowledge and (sometimes) innovations which may foster solutions to the problem of CC, but the sector is unable to legislate or to set binding regulations to steer public behaviour, as only political decision-makers can. HEIs, which in many countries are facing financial challenges due to shrinking demographics and reduced public funding (Jongbloed, 2022) as well as increasing competition (Krücken, 2021; Alajoutsijärvi et al., 2024), do not have the required resources to invest in green technologies and innovations on a broader scale, like the business or the investor sectors. There have been cases when hydrogen (zero carbon manufacturing) projects have come to a halt as the considerable amount of funding needed (and the business prospects) were eventually lacking (Kekäle & Pinheiro, 2024). Carbon dioxide capture, hydrogen, wind and solar energy, nuclear energy, electric vehicles—they all require funding and commercial and business solutions to proceed; in many such cases the scientific background of the innovations and technological applications are already mature and were established decades ago. When it comes to free choice in CC, much depends on consumption and fads among the public at large—the eight billion people on earth. The international dimension—different countries can diverge in diverse directions—is crucial as the CC problem is global in nature and adds to the complexity of the issue. This is one of the reasons why the book editors decided to study different country cases on an international basis, as to take stock of policy developments across multiple geographies, types of HEIs, and time scales.

Key Lessons from the Country Cases

In this volume, authors from a wide variety of countries and world regions undertook a systematic investigation of national HE policies and institutional responses with respect to HE and CC mitigation. The volume featured a total of nine country cases: Brazil, China, the Czech Republic, the Netherlands, Italy, Ireland, Poland, Switzerland, and the UK. The country studies focused especially on the nexus between national HE policy and the responses and activities of HEIs and their internal actors, sometimes in the absence of such policy framework. In that sense, the country cases provide a partial understanding, but they nonetheless reveal important aspects of the realities underpinning the national/local/global social systems in which CC mitigation occurs, both in rhetoric (symbolic and discursive) and practical (legislative, financial, and strategic) terms. Below, we present some of the key insights emanating from the cases.

Autonomous Institutions and Loose Political Steering

We should first note that there can be differences between overall national philosophies in thinking about CC mitigation. Kekäle and colleagues (Chap. 2) demonstrated that the overall approach towards CC mitigation and HEIs' *expected* roles differed in the UK and in the Netherlands. To simplify: the official UK national CC mitigation philosophy for HEIs' expected roles appears to be based on scientific analyses, by which legislation and legitimate political accountability is directed based on each sector's emissions. The parliamentary decisions are aided by an intermediating scientific body (CCC). In contrast, the Dutch national CC policy has initially been very collegial, based on national discussions and joint agreements and efforts. Consequently, Dutch HEIs are also expected to act more as CC policy implementors. Although the overall expectation or "pact" concerning HEIs' roles in CC mitigation appear to differ between the countries, in both cases the concrete actions have been relegated to HEIs themselves.

Regardless of such differences in the overall setup and construction of national CC policies, the overall impression based on the empirical cases—with the exception of China (Chap. 8)—is that the actual national *political steering* of HEIs in CC matters remains rather vague. What is more, the steering targets for HEIs tend to be continuously changing, cumulating, and cascading (Chap. 1). Political steering brings in new aims and targets on top of the previous ones as new parties take over. The outcome tends to be a complex and confusing set of expectations for HEIs, as pointed out in Chap. 1. The national steering mechanisms neither necessarily follow the changing political narratives nor are they likely to deviate from the policy discussions at the EU level. As an outcome, CC is an internationally recognised problem that does not fully penetrate national political steering mechanisms.

Confusion, compounded by the general lack of state steering (scarcity of funding and incentive systems targeting HEIs towards CC mitigation, and its key internal actors), appears to be the norm. In Poland (Chap. 6), the Ministry responsible for the HE sector has not provided any specific policy agenda to address grand challenges: tackling CC is a "bottom-up process guided by the interests of individual academics with little institutional engagement and rather passive policy measures" (Antonowicz et al., 2025, p. 159). Similarly, Kohoutek (Chap. 7) concludes that CC is not a strategic priority for HE policymaking in the Czech Republic, with the author speculating that this may have "something to do with a specific organizational response that sees climate change simply as just another agenda to be coped with under a condition of HEIs (and academics) constantly facing a capacity and mission overload" (Kohoutek, 2024, p. 188). Gibson and Hazelkorn (Chap. 3) echo these findings by noting that Irish HEIs are largely left to their own devices, resulting in considerable variety across and within universities concerning CC mitigation. In the case of Italy, Reale and Spinello (Chap. 4) point out that research policies have offered very limited funding opportunities to HEIs related to CC research, typically only supporting short-term projects. In addition, the authors argue that the country's CC policies lack internal alignment and that there is no binding national legislation. The same goes for the UK, where the government does not provide any concrete guidelines on how public bodies (such as HEIs) should meet their legally binding CC mitigation targets (Chaps. 2 and 5).

Beyond Europe, Axel-Berg and colleagues (Chap. 9) note that the Brazilian HE system is primarily designed for teaching, with funding often allocated to temporary independent research projects. This has resulted in a fragmented research structure where the autonomy of small research units remains a fundamental principle. In such situations, state-level funding bodies play an important role in bringing together a disjointed research system in the form of scientific (network-centred) programmes dedicated to addressing CC issues across the board. Finally, turning to China (Chap. 8), the account provided suggests that efforts by central government have been made in the last 15 years. Despite this, the country continues to be the single largest contributor of carbon emissions, attesting to its commitment towards economic growth and prosperity. HEIs are relatively newcomers as key stakeholders (2021 Action Plan), whose primary role is "to offer technological assistance and develop skilled individuals to achieve carbon peaking and neutrality ... a proactive measure that aligns with the national strategy" (Yu et al., 2025, p. 201, Chap. 8 in this volume). In this respect, China can be considered an *outlier* in the context of the other case studies included in this volume. This, however, is not surprising if one considers the top-down orientation underpinning the central state's role in the economy and society more generally, where considerable efforts have been made to maintain social (and party) cohesion sometimes at the expense of economic imperatives.

> China's climate change initiatives are part of a broader effort to enhance domestic and international legitimacy. The issue of climate change has become highly politicised in China under the leadership of President Xi Jinping, and it serves as a crucial test of the ruling party and the state's capacity to govern effectively. As such, *legitimacy-seeking is the key driver* that links China's domestic measures and international commitments. (Lian & Li, 2023, p. 171; emphasis added)

All in all, if one excludes the unique case of China and all its internal complexities, the overall findings prompt us to ask: *Why is it common to have such loose national political steering when it is almost universally accepted that the problem of CC is both urgent and severe?* A certain paradox can be observed. Globally, there is a clear emphasis on CC mitigation in the context of supranational HE policies (Kaldeway, 2018). Here, the best example is the European Union, with its Horizon 2020 flagship programme targeting research that directly contributes to achieving climate goals. The United Nations Sustainable Development Plan and its 17 strategic goals (SDGs) are being used as a global reference framework for HEIs across the globe. In the absence of a national policy framework, current strategic efforts tend to be based on independent HEIs, who are accountable public actors and receive funding on a political basis as well as based on their quantifiable results. State steering through legislation and resource allocations remains rather weak, even if there has been a shift towards a more salient top-down steering system in many national systems through the introduction (last 30 years or so) of performance management (Hazelkorn et al., 2018; Pinheiro et al., 2019).

The lack of national political steering is likely attributed to multiple political pressures and simultaneous crises affecting governments across the world (Trondal et al. 2022). The problem of CC is laden with complexity and ambiguity, as there is no silver bullet or one size fits all, requiring the adoption of a wide variety of solutions across the economy, polity, and society at large (Tertzakian & Hollihan, 2009). Better welfare, higher standards of living, and growing emissions have been two sides of the same coin. In political terms, the wicked problem of CC calls to mind Jean-Claude Juncker's[2] famous notion of knowing what to do, but not knowing how to get re-elected once we have done it (Juncker, 2007).

[2] Former Prime Minister of Luxembourg and 12th President of the European Commission (2014–2019).

Hence, political documents tend to avoid telling actors what is to be done, but rather respond to demands for *something to be done* (Chap. 3). Symbolic aspects underpinning political decision-making take central stage (March & Olsen, 1984), and any short term and politically dangerous initiatives—such as constraints on consumption or carbon taxation, for example—tend to be avoided. Nations compete for economic growth and a better standard of living, and citizens seem to request that too. In a globally integrated market economy, the room to manoeuvre, enabling any single government to make quicker and assertive decisions by themselves, might be counterproductive in economic terms and in the broader context of CC as a global *wicked problem* (Head, 2022), thus requiring an articulate response across the board.[3] This inability to act, in turn, tends to cause a vicious cycle, hindering CC mitigation.

International agreements, which could serve as a remedy, are slow and difficult to reach. Altogether, 196 countries (all with their own circumstances and strategic interests) were able to agree on the Paris COP (21) agreement in 2015, but it took 23 years to achieve this goal and some of the signatories have backtracked on their ambitious targets (Hodgkins, 2021). Gibson and Hazelkorn's (Chap. 3) account shows the difficulty of reaching national political decisions, regardless of (Irish) people's apparently strong support for CC mitigation. Political opposition attacks, rising populism, and the potential worsening of living conditions might easily change the domestic political atmosphere against the cuts. Policy makers' desires often became "all things to all audiences".

Cooperation between political decisionmakers and HEIs is multifaceted (Capano & Jarvis, 2020) and, apparently, problematic in terms of CC mitigation as well. There is evidence suggesting that HEIs are not necessarily doing the utmost in fostering CC mitigation, in part due to cultural factors that are context dependent. Historically, in many countries, such as Poland, HEIs have not been expected to be engaged in addressing social and economic problems as reported by Antonowicz and colleagues (Chap. 6). Academic autonomy and freedom, core values in

[3] Economically more viable, scalable, and transferable global (energy) solutions for transport, e.g., would help any nation or businesses in international competition by bringing any global player an advantage. Progress in CC mitigation has been made, but "although wind and solar are now cheaper than fossil fuels" in electricity production, elsewhere easy, novel solutions are not currently at hand as "virtually every human activity produces greenhouse gas emissions" and "some of the clean technologies we need are still very far from becoming practical, cost-effective solutions we can deploy at scale" (Gates, 2022, pp. 1–2).

academia, are cemented in legislation, and can act as barriers for enacting stronger state-centred policies such as CC mitigation, as is also the case in Brazil (Chap. 9). Indeed, it is difficult to imagine any good research being done without strong voluntary and intrinsic commitment by the academic heartland (Clark, 2001). If the political steering is difficult, the steering within HEIs themselves is additionally troublesome with internal actors disagreeing on strategic goals and the means to achieve them (Kekäle & Pinheiro, 2025). But a logical side-effect of this free and individualistic culture with very little political or institutional steering is that individual researchers are the *key agents* in deciding what to study and whether to engage with CC-related research, as outlined in the Polish case (Chap. 6). Thus, the quality of academics and success in their recruitment becomes the fundamental issue (Shattock, 2010). Yet, one could claim that a strong institutional commitment to the principles of sustainability is essential if CC mitigation is to be embedded into HEIs' structures, core activities, and academic practices (Cristofoletti & Pinheiro, 2023). Below, we take stock of how the situation appears empirically across the cases composing the bulk of this edited volume.

HEIs' Measures Towards CC Mitigation

Many contributions in the volume provide descriptions of institutional actions in relation to CC mitigation. In this respect, it is worth revisiting Charles' (2024) account (Chap. 5 in this volume) on the ways in which HEIs generally contribute to sustainability more broadly, namely:

1. *Via institutional and estate management practices (affecting the emissions they produce);*
2. *As sources of research and expertise in the technical and holistic aspects of sustainable development;*
3. *Educational and cultural roles* in providing a common educational experience for students;
4. *Active involvement with activities directed towards local and regional sustainability.*

In many countries, as attested by some of the case studies in this volume, HEIs have been central to the creation of a knowledge or expertise base (point 2 above) which has informed government policies towards CC mitigation. The other contributions referred to by Charles (Chap. 5) are also noteworthy, but given the insurmountable global problem at hand, HEIs are mainly undertaking more of a local role in CC mitigation at present, providing the potential for broader solutions. Research or knowledge production is one of HEIs' key tasks, in the Humboldtian tradition, and it is seen as the basis for knowledge transmission or teaching (Nybom, 2003). The overall picture provided by the rich empirical accounts in this volume is shaped as such: Individual researchers' and teams' interests and commitments ***mainly drive research*** in CC, and sufficiently loose, national, curiosity-based funding programmes, plus the EU funding agendas, ***allow*** such research to be carried out across disciplines. Institutional strategies and priorities appear to be relatively loosely coupled and highly symbolic for the most part, a feature that resembles typical postures within HEIs from a historical perspective (Young & Pinheiro, 2022). Most of the empirical cases do not report strong institutional involvement with respect to dedicated programmes and practical orientations for dealing with CC. This relatively weak and symbolic institutional stance is somewhat surprising as a pattern of strong institutional leadership has been increasingly observed in many countries (Kekäle & Pinheiro, 2025), including external pressures for leveraging HEIs' societal contributions or impact (Avramovic et al., 2022).

The strongest case illustrating this point (a strong emphasis on individual researchers combined with weak institutions) can be found in national systems like Brazil's (Chap. 9). Project grants and scholarships—governmental but also from other funders—have been the sole instruments for supporting research activities inside Brazilian universities. Thus, there is direct external support available to researchers through projects and only a minor involvement by HEIs themselves in setting research agendas. All budgeting and expenditures are linked to the research projects in question. Deliverables mostly consist of published material, with research publications being produced in great numbers. However, in the Amazonian region, the case found that the project's tasks can be tightly connected to practical outcomes in sustainable development, such as protecting natural habitats and re-vitalising local occupations such as fishing.

Similarly, in their analysis of doctoral dissertations at Italian universities as a proxy for new research in CC, Reale and Spinello (Chap. 4) show how, as in Brazil (Chap. 9), CC research is mostly left to the free initiative of the universities and scholars, based on project funding. Yet the numbers are compelling: between 2010 and 2019, a total of 1036 Italian theses addressed CC topics. However, the data show a general upward trend in the numbers of CC-related dissertations until 2017, and, for some unexplained reason, a significant decrease after that. One could speculate whether the saliency of national populist parties and the enactment of grand coalitions (avoiding important but controversial issues, like CC) since the 2018 (Italian) election may have something to do with this.

In Chap. 6, Antonowicz and colleagues point out that Polish HEIs are publicly funded and enjoy high levels of institutional autonomy, as is the case in Brazil. As noted, the national policy steering is loose, mainly following European Green Policy in a vague manner, and there appears to be a clear political unwillingness to acknowledge the seriousness of CC. Unsurprisingly, HEIs' strategies appear to be aligned with the governmental posture, being mostly ceremonial (symbolic compliance) rather than representing any real engagement or commitment. Yet again, as in Italy, on an individual level, there has been a rapid and exponential rise in CC-related research. In 2022, almost one-third of projects (around 100) funded by the Polish National Research Council were related to CC. The writers echo the overall storyline by assuming that this is mainly a result of the research interests of individual scientists, as most HEIs' strategies include CC without really specifying what this stands for. With that said, Antonowicz et al. conclude, on an optimistic note, in hoping that this strong bottom-up academic movement will, eventually, boost institutional engagement with CC and will also impact public policies in the mid-term.

Moving to Ireland, Gibson and Hazelkorn (Chap. 3) discuss institutional and individual actions in relation to CC. They echo the view expressed in many other chapters, namely: although the national steering is very loose, and funding instruments tend to focus mainly on "developing economically strategic areas" for "drawing prestigious international scientists into Ireland" and "human capital development", there has, nonetheless, been enough room for curiosity-driven approaches, enabling CC research to be funded at the individual level (Gibson & Hazelkorn, 2025). The authors contend that this "tells us something about the priorities of researchers applying for funding", especially for career-specialising PhDs and early career researchers with no tenure, but this also highlights the priorities of influential members of funding bodies (many of them academics themselves) who select "the topics they believe are worth funding" (Gibson & Hazelkorn, 2025, p. 78).

Beyond research, Kohoutek's account of dynamics within the Czech Republic (Chap. 7) refers to the role of HEIs' educational activities in shaping human mindsets (norms and values), as well as their broader societal mission in raising awareness towards CC and in promoting sustainable strategies and/or programmes targeting multiple audiences. However, the author notes that regardless of their significant potential, HEIs' overall role in CC issues seems rather ambiguous: the impact is long-term and nebulous, and politicians are often reserved in establishing binding commitments based on scientific findings. Moreover, HEIs themselves (in the roles of research, education, and societal cooperation) do not only try to mitigate CC, but also often foster the exploitation of natural resources and production based on fossil fuels. This process may be dependent on values, orientation, and disciplinary bases. Collectively, HEIs directly produce emissions through their staff's international travel practices—a problem which has not been solved but it is starting to be addressed somewhat by some HEIs, as in the UK (Charles, Chap. 5).

Finally, Felder-Stindt's account (Chap. 10) identifies 27 alliances that address issues of CC and sustainability. Due to political steering and funding, it is not surprising that these alliances' mission statements "strongly reflect EU speech" as required deliverables of the alliances (Felder, 2025, p. 251). The author concludes that HEIs' alliances feature CC and sustainability topics in their missions because of the profiles of the partners involved (norms and values) and due to societal demands and funding stipulations (compliance and strategic considerations). Cooperation results are primarily in terms of incorporating CC mitigation in delivering short-term joint teaching formats and in researchers working together on similar topics and shared agendas. Alliances have also served to exchange best practices. The establishment of joint study programmes proves to be more complex because of diverging HE systems and regulations. In short, HEIs are increasingly responsive to societal and political developments, and funding prospects, but these actions also serve to position HEIs both politically (legitimacy) and strategically (global competition), while also helping to contribute to the fight against CC.

Conclusions

HEIs' role in CC mitigation is part and parcel of a larger system of influences involving a wide variety of key players at multiple levels and jurisdictions (Fig. 1.1, Chap. 1). The rich empirical accounts across the cases provided here suggest that HEIs' roles, regarding societal outcomes, in CC mitigation are bound to political decision-making on the one hand, and to the industrial and social usage (social impact) of the knowledge produced by researchers on the other. Despite their multiple competencies, HEIs do not, by themselves, have the power or capacity to mobilise system-wide changes on their own, hence they are dependent on larger institutional frameworks which underpin their social role (Pinheiro et al., 2023).

Politics and symbolic actions were found to play a legitimising role in CC mitigation. This applies to political steering and to HEIs' strategies which often stress ideological and symbolic dimensions with high expectations and unrealistic promises (Enders & Boer, 2009). It is abundantly clear that individual solutions only go so far and that there is a need for collective and coordinated policy and institutional responses (at multiple scales and levels) to the CC problem. Yet, the global problem of CC—as a long-term wicked policy dilemma—is disproportionally related to the short funding periods and the daily competitive realities facing most HEIs worldwide. There is still much to be done by HEIs alone.

If HEIs are willing to strengthen their impact in CC mitigation and leave political and symbolic "greenwashing activities" aside (which do not help in CC mitigation), they may be able to focus on specific inner capabilities and strengths at their disposal, as recently argued by Stensaker and Hermansen (2025). In most cases, institutional leadership within HEIs has the potential to help organise strong units around strategic CC areas, to foster policies for the recruitment of motivated researchers, and to find creative solutions (together with societal actors across the public, private, and civic sectors) to finance key strategic priorities regarding CC mitigation. Given the considerable complexity of the university as an organisation and fiduciary institution (Olsen, 2007; Pinheiro et al., 2018), it is imperative that policy makers and institutional leaders move away from "one size fits all approaches", and instead, devise policies, strategies, and incentives that consider the complex nature of HEIs and the global knowledge/science systems in which they are more generally embedded.

On the technical (STEM-related) side of the problem of CC, potential strategic targets might encompass diverse solutions to decarbonise economies and energy production, as advocated by leading public figures such as Bill Gates (Gates, 2021). Another strategic area of interest related to CC would be "carbon capture, utilisation and storage (CCUS) technology" which should be "an area of priority research interest" (House of Commons, 2022, pp. 71–75). New clean technologies by HEIs and companies would be needed, as would additional global political decisions to drive down the cost of the new clean technologies (and to make the emitting ones more expensive) to deploy them in the markets (Gates, 2022).

Citizens' consumption and behaviour are a fundamental part of the overall global emissions build-up. Human behaviour is the traditional realm of the social sciences and the humanities, with the potential, among other things, to offer alternative conceptions of the "good life", combat consumerism, and alternatives to a growth-obsessed and energy intensive economy (Selby & Kagawa, 2010, pp. 42–44). As socialisers of the next generation of citizens and consumers, HEIs have a strategic opportunity to rethink and redesign curriculum structures that consider CC mitigation. Equality-related dimensions (gender, ethnicity, social background, etc.) are also relevant aspects for HEIs regarding CC-related teaching and research. Legal complications in CC legislation might also need to be addressed as well, but this wide range of issues goes beyond the scope of this volume and could be addressed in future research inquiries.

Finally, given the interrelated nature of the problem (as depicted in Figs. 1.1 and 1.2, Chap. 1), HEIs' strategic roles in CC mitigation may best be enacted by providing vital knowledge and analytical insights to decision-makers at local, national, regional, and global levels; in addition to working alongside actors across the public, private, and civil society sectors in helping to: provide co-creative solutions for green technologies, solving bottlenecks for business processes associated with green transitions, developing coping and adjustment strategies, and fostering a CC-related cultural mindset by socialising students and re-educating other citizens alike. The future will reveal whether HEIs—together with other key societal actors—are up to this critical yet daunting task.

References

Alajoutsijärvi, K., Kettunen, K., & Pinheiro, R. (2024). Ideologies shaping university competition. *Studies in Higher Education, 49*(10), 1710–1719. https://doi.org/10.1080/03075079.2024.2397048

Antonowicz, D., Wróblewski, M., Suchomska, J. & Budzanowska, A. (2025). Mapping efforts in Polish Higher Education to tackle Climate Change. In J. Kekäle & R. Pinheiro (Eds.), *Higher Education Policy for Tackling Climate Change: Drivers, Dynamics and Effects* (pp. 141–169). Palgrave.

Avramovic, A., Pinheiro, R., & Asante, M. (2022). To be or not to be responsible? Academic research and social responsibility. *Higher Education Forum, 19*, 131–153.

Back to Earth (2024) Bring knowledge back to Earth. Let's come together and demonstrate the impact that research has on the real world. Bring knowledge back to Earth: Academic Evaluation and Impact (elsevier.com) Online. Retrieved March 6, 2024.

Capano, G., & Jarvis, D. S. (2020). *Convergence and diversity in the governance of higher education: Comparative perspectives*. Cambridge University Press.

Clark, B. (2001). The Entrepreneurial University: New foundations for collegiality, autonomy, and achievement. *Higher Education Management, 13*(2), 9–24.

Cristofoletti, E. C., & Pinheiro, R. (2023). Greening the University?: Assessing the impact of sustainability and SDGs in universities' values and strategies. In I. W. Filho, F. Frankenberger, & U. Tortato (Eds.), *Sustainability in practice: Addressing challenges and creating opportunities in Latin America* (pp. 111–126). Springer.

Enders, J., & Boer, H. (2009). The mission impossible of the European University: Institutional confusion and institutional diversity. In A. Amaral, G. Neave, C. Musselin, & P. Maassen (Eds.), *European integration and the governance of higher education and research* (Vol. 26, pp. 159–178). Springer.

Gates, B. (2021). *How to avoid climate disaster*. Vintage Books.

Gates, B. (2022). The state of the energy transition. A Blog. October 18, 2022. Online, Retrieved September 2, 2024, from https://www.gatesnotes.com/2022-State-of-the-Energy-Transition

Gibson, A. & Hazelkorn, E. (2025). Reframing, reacting, or responding? The climate emergency and higher education in Ireland. In J. Kekäle & R. Pinheiro (Eds.), *Higher Education Policy for Tackling Climate Change: Drivers, Dynamics and Effects* (pp. 63–88). Palgrave.

Hazelkorn, E., Coates, H., & McCormick, A. C. (2018). *Research handbook on quality, performance and accountability in higher education*. Edward Elgar Publishing.

Head, B. W. (2022). *Wicked problems in public policy*. Springer.

Hodgkins, R. (2021). COP26: A four-minute guide by a climate scientist. Loughborough University. Online, Retrieved March 7, 2024, from https://www.lboro.ac.uk/news-events/news/2021/november/cop-guide-conversation-richard-hodgkins/

House of Commons. (2022). Science and Technology Committee. The role of hydrogen in achieving Net Zero Fourth Report of Session 2022–23. Retrieved March 29, 2023, from https://committees.parliament.uk/publications/33292/documents/180198/default/

Jongbloed, B. (2022). The funding of higher education: International perspectives. In R. Tierney, F. Rizvi, & K. Erkican (Eds.), *International encyclopedia of education* (4th ed., pp. 183–190). Elsevier. https://doi.org/10.1016/B978-0-12-818630-5.02047-9

Juncker, J-C (2007, March 15). The quest for prosperity. *The Economist*.

Kagawa, F., & Selby, D. (Eds.). (2010). *Education and climate change: Living and learning in interesting times* (Vol. 30). Routledge.

Kaldeway, D. (2018). The grand challenges discourse: Transforming identity work in science and science policy. *Minerva, 56*(2), 161–182.

Kekäle, J., & Pinheiro, R. (2024). Steering regional green transitions: Assessing the role of knowledge, politics, and markets. A paper presented in HEREGREEN workshop: *The Role of Nordic Higher Education in Regional Green Transitions.* Tampere, 18–19 April 2024.

Kekäle, J., & Pinheiro, R. (2025). Individual Leadership in Higher Education: Partial solutions, unintended consequences, and the search for an alternative. In B. Rao (Ed.), *Worldviews and values book series—Volume II: Institutional governance, leadership, capacity building* (pp. 137–149). Emerald.

Krücken, G. (2021). Multiple competitions in higher education: A conceptual approach. *Innovation, 23*(2), 163–181.

Lian, C., & Li, J. (2023). Legitimacy-seeking: China's statements and actions on combating climate change. *Third World Quarterly, 45*(1), 171–188.

March, J., & Olsen, J. (1984). The new institutionalism: Organizational factors in political life. *The American Political Science Review, 78*(3), 734–749.

Nybom, T. (2003). The Humboldt Legacy: Reflections on the past, present, and future of the European University. *Higher Education Policy, 16*(2), 141–159.

Olsen, J. P. (2007). The institutional dynamics of the European university. In P. Maassen & J. P. Olsen (Eds.), *University dynamics and European integration* (pp. 25–54). Springer.

Pinheiro, R., Balbachevsky, E., Pillay, P., & Yonezawa, A. (2023). *The impact of Covid19 on the institutional fabric of higher education: Accelerating old patterns, imposing new dynamics.* Palgrave.

Pinheiro, R., Geschwind, L., Hansen, H. F., & Pulkkinen, K. (2019). *Reforms, organizational change and performance in higher education: A comparative account from the Nordic countries.* Palgrave.

Pinheiro, R., Šima, K., Young, M., & Kohoutek, J. (2018). University complexity and regional development in the periphery. In R. Pinheiro, M. Young, & K. Sima (Eds.), *Higher education and regional development* (pp. 1–20). Springer.

Reale, E., & Spinello, A. (2024). Government R&D funding policy for academic research: Are there incentives for climate change solutions? In: Kekäle, J. & Pinheiro, R. (Eds.), *Higher education policy for tackling climate change* (pp. XX–XX). Palgrave Macmillan.

Shattock, M. (2010). *Managing successful universities.* SHRE & Open University Press.

Spangenberg, J. H. (2017). Hot air or comprehensive progress? A critical assessment of the SDGs. *Sustainable Development, 25*(4), 311–321.

Stensaker, B., & Hermansen, H. (2025). Global, Nordic, or institutional visions? An investigation into how Nordic universities are adapting to the SDGs. *Higher Education, 89*, 293–309. https://link.springer.com/article/10.1007/s10734-023-01047-3

Tertzakian & Hollihan. (2009). *The end of energy obesity: Breaking today's energy addiction for a prosperous and secure tomorrow.* Wiley.

Trondal, J., Keast, R., Noble, D., & Pinheiro, R. (2022). *Governing complexity in times of turbulence*. Edward Elgar.

Young, M., & Pinheiro, R. (2022). The Post-entrepreneurial university: The case for resilience in higher education. In R. Pinheiro, M. L. Frigotto, & M. Young (Eds.), *Towards resilient organizations and societies: A cross-sectoral and multi-disciplinary perspective* (pp. 173–193). Palgrave.

Yu, H., Linlin, T. & Botao, H. (2025). Responses to Climate Change: Context, Action, and Future Directions of China's Universities. In J. Kekäle & R. Pinheiro (Eds.), *Higher Education Policy for Tackling Climate Change: Drivers, Dynamics and Effects* (pp. 197–216). Palgrave.

Index[1]

[1] Note: Page numbers followed by 'n' refer to notes.

J. Kekäle, R. Pinheiro (eds.), *Higher Education Policy for Tackling Climate Change*, https://doi.org/10.1007/978-3-031-78479-8

The manufacturer's authorised representative in the EU is Springer Nature Customer Service Centre GmbH, Europaplatz 3, 69115 Heidelberg, Germany. If you have any concerns regarding our products, please contact ProductSafety@springernature.com

Printed and bound by CPI Group (UK) Ltd, Croydon, CR0 4YY
30/09/2026
02225595-0007